Methods in Molecular Biology™

Series Editor
John M. Walker
School of Life Sciences
University of Hertfordshire
Hatfield, Hertfordshire, AL10 9AB, UK

For further volumes:
http://www.springer.com/series/7651

The Inflammasome

Methods and Protocols

Edited by

Christine M. De Nardo and Eicke Latz

Institute of Innate Immunity, University of Bonn, Bonn, Germany

Humana Press

Editors
Christine M. De Nardo, PhD
Institute of Innate Immunity
University of Bonn
Bonn, Germany

Eicke Latz, MD, PhD
Institute of Innate Immunity
University of Bonn
Bonn, Germany

ISSN 1064-3745 ISSN 1940-6029 (electronic)
ISBN 978-1-4939-6011-8 ISBN 978-1-62703-523-1 (eBook)
DOI 10.1007/978-1-62703-523-1
Springer New York Heidelberg Dordrecht London

Printed on acid-free paper

Humana Press is a brand of Springer
Springer is part of Springer Science+Business Media (www.springer.com)

Preface

Cells of the immune system, such as macrophages and dendritic cells, continuously survey host tissue integrity. Immune cells can quickly respond to invading pathogens or tissue damage by mounting an immune response geared to the reconstitution of tissue homeostasis. Immune cells have the ability to distinguish foreign molecules from self-molecules, detect altered self-molecules, or respond to metabolic derangements via the expression of families of signaling receptors that can recognize such danger signals. Most of these immune signaling receptor families, including Toll-like receptors, Rig-I like receptors or C-type lectin receptors, as well as a number of Nod-like receptors (NLRs), induce signaling cascades that eventually culminate in a transcriptional response of the cell and the production of inflammatory mediators. The activated factors act in concert with many cell types that help in the defense against the invading pathogen or lead to the repair of damaged tissues.

Inflammasomes, which are the main topic of this volume, have a seemingly simple architecture. A sensor molecule of the NLR or PYHIN protein family recruits the common adapter molecule apoptosis-associated speck-like protein containing a CARD (ASC), which then leads to the approximation of pro-caspase-1 molecules inducing their autoproteolytic activation. Intriguingly, following activation, ASC forms a large protein aggregate, termed "ASC speck," which is thought to provide a platform for the activation of caspase-1. Active caspase-1, in turn, cleaves the inactive precursor molecules of the IL-β cytokine family (IL-1β and IL-18) into their biologically active forms. In addition, inflammasomes control the release of mature IL-1β cytokine family members, demonstrating that inflammasomes are key gatekeepers for these highly pro-inflammatory cytokines. Indeed, uncontrolled production and activation of IL-1β cytokines can lead to significant inflammatory reactions that contribute to a range of diseases. For example, the NLRP3 inflammasome, which recognizes a range of microbes as well as many sterile danger signals, can contribute to common inflammatory pathologies, including gout, atherosclerosis, type 2 diabetes, and Alzheimer's disease.

While we have learned a great deal about the mechanisms leading to the production of IL-1β family cytokines, more precise details of how inflammasomes are activated, remain to be elucidated. Elaborate mechanisms have evolved that control the activity of inflammasomes and we are just beginning to understand the upstream mechanisms that lead to the formation of inflammasomes by the many reported triggers. Future work in this area will reveal novel targets for pharmacological interference and could lead to more specific anti-inflammatory interventions that are urgently needed.

Protocols used in the study of inflammasomes can be difficult and intricate to master in the beginning and thus, detailed protocols with tips from the experts can be of great value. A number of expert labs have developed specific techniques to study many aspects of inflammasome function in human or mouse immune cells. Furthermore, some benchmark protocols, such as the analysis of caspase-1 cleavage by immunoblotting, have evolved individually in different labs around the world. We have the unique opportunity to present these protocols in one volume, which will help broaden the inflammasome field by allowing others to more easily and hopefully more successfully perform these assays.

Protocols are provided that detail the generation of inflammasome stimuli, such as the NLRP3-stimulating amyloid beta and islet amyloid amylin peptides. Other protocols describe how perturbance of cellular homeostasis by crystalline materials, infectious agents, or certain adjuvants can be used to activate inflammasomes. Inflammasome activation can be studied at several levels, hence, detailed protocols including the assessment of ASC speck formation, the monitoring of caspase-1 activity and processing, and the activation of IL-1β cytokines are presented. In addition, protocols for the analysis of inflammasome assembly and ATP binding or ATPase activity of inflammasomes are described. Another hallmark of inflammasome activation is the induction of a specialized form of cell death called "pyroptosis." Methods to quantify this process as well as the assessment of consequences of inflammasome activation for the cell and host are also detailed in this volume. These chapters will be a useful resource for investigators who seek a better understanding of inflammasome activation pathways and would like to master techniques optimized by experts in the field. We hope this collection will form the definite lab protocol source for inflammasome research.

This project would not have been possible without the help of some key people. Thank you to Dr. John Walker and David Casey for identifying a niche and providing support along the way. We are also very grateful to Dr. Dominic De Nardo for critically reviewing all the manuscripts. Most importantly, we would like to thank all the authors for their excellent contributions, enthusiasm, and kindness throughout this project. These protocols and the personal tips and notes shared by each author will form an invaluable tool for current and future generations of scientists interested in inflammation and inflammasome research.

Bonn, Germany

Christine M. De Nardo
Eicke Latz

Contents

Contributors

EVA BARTOK • *Institute for Clinical Chemistry and Pharmacology, Unit for Clinical Biochemistry, University Hospital, University of Bonn, Bonn, Germany*

FRANZ BAUERNFEIND • *Institute for Clinical Chemistry and Pharmacology, Unit for Clinical Biochemistry, University Hospital, University of Bonn, Bonn, Germany*

HANNES BECKERT • *Max-Planck Research Group Neuroimmunology, Center of Advanced European Studies and Research (caesar), Bonn, Germany*

PETR BROZ • *Department of Microbiology and Immunology, Stanford School of Medicine, Stanford University, Stanford, CA, USA*

VISHWA DEEP DIXIT • *Immunobiology Laboratory, Pennington Biomedical Research Center, Louisiana State University System, Baton Rouge, LA, USA*

PETER DUEWELL • *Medizinische Klinik und Poliklinik IV, Klinikum der Universität München, Munich, Germany*

JOSEPH A. DUNCAN • *Division of Infectious Diseases, Department of Medicine, University of North Carolina School of Medicine, Chapel Hill, NC, USA*

AISLING DUNNE • *Immunology Research Centre, School of Biochemistry and Immunology, Trinity College, Dublin 2, Ireland*

ERAN ELINAV • *Department of Immunobiology, Yale University School of Medicine, New Haven, CT, USA; Immunology Department, Weizmann Institute of Science, Rehovot, Israel*

BENJAMIN FAUSTIN • *Sandford-Burnham Institute for Medical Research, La Jolla, CA, USA*

RICHARD A. FLAVELL • *Department of Immunobiology, Yale University School of Medicine, New Haven, CT, USA; Howard Hughes Medical Institute, Chevy Chase, MD, USA*

RYAN GRANT • *Immunobiology Laboratory, Pennington Biomedical Research Center, Louisiana State University System, Baton Rouge, LA, USA*

OLAF GROß • *Institut für Klinische Chemie und Pathobiochemie, Klinikum rechts der Isar, Technische Universität München, Munich, Germany*

ANNETT HALLE • *Max-Planck Research Group Neuroimmunology, Center of Advanced European Studies and Research (caesar), Bonn, Germany*

CLAIRE H.A. HEARNDEN • *Adjuvant Research Group, School of Biochemistry and Immunology, Trinity Biomedical Sciences Institute, Trinity College, Dublin 2, Ireland*

VEIT HORNUNG • *Institute for Clinical Chemistry and Pharmacology, Unit for Clinical Biochemistry, University Hospital, University of Bonn, Bonn, Germany*

GABOR L. HORVATH • *Institute of Innate Immunity, University Hospital, University of Bonn, Bonn, Germany*

CHRISTOPHER JAKOBS • *Institute for Clinical Chemistry and Pharmacology, Unit for Clinical Biochemistry, University Hospital, University of Bonn, Bonn, Germany*

LEO A.B. JOOSTEN • *Department of Medicine, Radboud University Nijmegen Medical Center, Nijmegen, The Netherlands; Nijmegen Institute for Infection, Inflammation and Immunity (N4i), Nijmegen, The Netherlands*
MAN LYANG KIM • *Inflammation Division, The Walter and Eliza Hall Institute of Medical Research, Melbourne, VIC, Australia*
ERIC M. KOFOED • *Division of Immunology and Pathogenesis, Department of Molecular and Cell Biology, University of California, Berkeley, CA, USA*
HAJIME KONO • *Department of Internal Medicine, Teikyo University School of Medicine, Tokyo, Japan*
ANDREJ KUBARENKO • *Institute for Clinical Chemistry and Pharmacology, Unit for Clinical Biochemistry, University Hospital, University of Bonn, Bonn, Germany*
EICKE LATZ • *Institute of Innate Immunity, University Hospital, University of Bonn, Bonn, Germany; Division of Infectious Diseases and Immunology, Department of Medicine University of Massachusetts Medical School, Worcester, MA, USA; German Center for Neurodegenerative Diseases (DZNE), Bonn, Germany*
ED C. LAVELLE • *Adjuvant Research Group, School of Biochemistry and Immunology, Trinity Biomedical Sciences Institute, Trinity College, Dublin 2, Ireland*
SETH L. MASTERS • *Inflammation Division, The Walter and Eliza Hall Institute of Medical Research, Melbourne, VIC, Australia*
EDWARD A. MIAO • *Department of Microbiology and Immunology, University of North Carolina at Chapel Hill, Chapel Hill, NC, USA*
JINYAO MO • *Division of Infectious Diseases, Department of Medicine, University of North Carolina School of Medicine, Chapel Hill, NC, USA*
DENISE M. MONACK • *Department of Microbiology and Immunology, Stanford School of Medicine, Stanford University, Stanford, CA, USA*
BRIAN G. MONKS • *Institute of Innate Immunity, University Hospital, University of Bonn, Bonn, Germany; Department of Medicine, University of Massachusetts Medical School, Worcester, MA, USA*
MIHAI G. NETEA • *Department of Medicine, Radboud University Nijmegen Medical Center, Nijmegen, The Netherlands; Nijmegen Institute for Infection, Inflammation and Immunity (N4i), Nijmegen, The Netherlands*
EWA OLESZYCKA • *Adjuvant Research Group, School of Biochemistry and Immunology, Trinity Biomedical Sciences Institute, Trinity College, Dublin 2, Ireland*
KAORU ORIHASHI • *Department of Internal Medicine, Teikyo University School of Medicine, Tokyo, Japan*
THEO S. PLANTINGA • *Department of Medicine, Radboud University Nijmegen Medical Center, Nijmegen, The Netherlands; Nijmegen Institute for Infection, Inflammation and Immunity (N4i), Nijmegen, The Netherlands*
ANTHONY RAVUSSIN • *Immunobiology Laboratory, Pennington Biomedical Research Center, Louisiana State University System, Baton Rouge, LA, USA*
MANIRA RAYAMAJHI • *Department of Microbiology and Immunology, University of North Carolina at Chapel Hill, Chapel Hill, NC, USA*
JOHN C. REED • *Sandford-Burnham Institute for Medical Research, La Jolla, CA, USA*
MAREIKE SCHNAARS • *Max-Planck Research Group Neuroimmunology, Center of Advanced European Studies and Research (caesar), Bonn, Germany*
KATHARINA S. SCHNEIDER • *Institut für Klinische Chemie und Pathobiochemie, Klinikum rechts der Isar, Technische Universität München, Munich, Germany*

ANDREA STUTZ • *Institute of Innate Immunity, University Hospital, University of Bonn, Bonn, Germany*
CHRISTOPH A. THAISS • *Department of Immunobiology, Yale University School of Medicine, New Haven, CT, USA; Institute of Microbiology, ETH Zurich, Zurich, Switzerland*
CHRISTINA J. THOMAS • *Institut für Klinische Chemie und Pathobiochemie, Klinikum rechts der Isar, Technische Universität München, Munich, Germany*
RUSSELL E. VANCE • *Division of Immunology and Pathogenesis, Department of Molecular and Cell Biology, University of California, Berkeley, CA, USA*
C. BRUCE VERCHERE • *Department of Pathology and Laboratory Medicine and the Department of Surgery, Child and Family Research Institute, University of British Columbia, Vancouver, BC, Canada*
CLARA WESTWELL-ROPER • *Department of Pathology and Laboratory Medicine, Child and Family Research Institute, University of British Columbia, Vancouver, BC, Canada*
TAMIKO YANAGIDA • *Department of Internal Medicine, Teikyo University School of Medicine, Tokyo, Japan*
MARIE YANG • *Adjuvant Research Group, School of Biochemistry and Immunology, Trinity Biomedical Sciences Institute, Trinity College, Dublin 2, Ireland*
YUN-HEE YOUM • *Immunobiology Laboratory, Pennington Biomedical Research Center, Louisiana State University System, Baton Rouge, LA, USA*
YUE ZHANG • *Department of Microbiology and Immunology, University of North Carolina at Chapel Hill, Chapel Hill, NC, USA*

Chapter 1

Assessing β-Amyloid-Induced NLRP3 Inflammasome Activation in Primary Microglia

Mareike Schnaars, Hannes Beckert, and Annett Halle

Abstract

Senile plaques are an important histological hallmark of Alzheimer's disease. They mainly consist of the fibrillar peptide β-amyloid (Aβ) and are surrounded by activated microglia and astrocytes. Microglia in the vicinity of senile plaques express high levels of proinflammatory cytokines and neurotoxic substances, which are believed to influence disease progression. One important cytokine in Alzheimer's disease is IL-1β. Stimulation of cultured primary microglia by synthetic fibrillar Aβ causes the release of IL-1β via activation of the NLRP3 inflammasome.

Here we provide protocols for the preparation of primary microglial cultures and synthetic oligomeric and fibrillar forms of Aβ.

Key words Inflammasome, Primary microglial cells, Oligomeric, Fibrillar, β-Amyloid, Aβ

1 Introduction

Alzheimer's disease is the most common form of dementia and is associated with progressive memory loss, cognitive decline and neuropsychiatric symptoms in affected patients. Two important histological hallmarks of the disease are senile plaques and neurofibrillary tangles, both of which are believed to play a major role in the initiation and progression of disease. Senile plaques are formed by aggregates of the extracellular fibrillar peptide, which is generated upon differential cleavage from the ubiquitously expressed type I transmembrane protein amyloid precursor protein (APP). Aggregation of Aβ monomers leads to the formation of soluble and insoluble Aβ species e.g., oligomers, protofibrils, fibrils, seeds, and plaques. The relevance of these different Aβ species, which are simultaneously present in Alzheimer's disease patients, is under investigation. It has become evident that oligomeric Aβ species are more neurotoxic when compared to fibrillar Aβ [1] and there is a correlation between the levels of oligomeric Aβ and cognitive

Christine M. De Nardo and Eicke Latz (eds.), *The Inflammasome: Methods and Protocols*, Methods in Molecular Biology, vol. 1040, DOI 10.1007/978-1-62703-523-1_1, © Springer Science+Business Media New York 2013

dysfunction in Alzheimer's disease patients [2]. Thus, it is very important to further dissect the molecular and cellular effects of specific Aβ species.

Microglia are the resident innate immune cells of the brain. They migrate to senile plaques in Alzheimer's disease patients and have been shown to phagocytose Aβ in vitro [3] and in vivo [4]. Aβ-plaque associated microglia express higher levels of IL-1β [5, 6] and elevated levels of IL-1β have been found in the cerebrospinal fluid (CSF) of Alzheimer's disease patients [7]. We have reported that uptake of fibrillar Aβ by primary microglia induces the release of mature IL-1β [8], which may provide an explanation for elevated IL-1β production in Alzheimer's disease patients. By using different assays to assess inflammasome activation, such as ELISAs, Western blotting for caspase-1 cleavage and cleavable fluorescence substrates as well as cells from inflammasome component-deficient mice, we could show that Aβ-induced IL-1β release was dependent on caspase-1 and NLRP3 inflammasome activation. Here we will describe protocols for the generation of primary microglial cells and the preparation of synthetic oligomeric and fibrillar Aβ, allowing the effect of different Aβ species to be studied.

2 Materials

2.1 Preparation and Cultivation of Primary Microglial Cells

1. Binocular microscope (7× to 45× magnification, min. 94 mm working distance, with external or integrated illumination (20 W)).
2. Ice and ice buckets.
3. Sterile serological pipettes.
4. Sterile micropipette tips.
5. Serological pipette and micropipette.
6. Inverse light microscope, phase-contrast equipment.
7. Hemocytometer.
8. Laminar flow hood.
9. Cell culture incubator maintained at 37 °C and 5 % CO_2.
10. Swinging bucket centrifuge.
11. Water bath, 37 °C.
12. Sterile, cell culture grade PBS, without Ca^{2+}, Mg^{2+}.
13. Cell culture medium: DMEM high glucose supplemented with L-glutamine, 10 % (v/v) FBS, 1 % (v/v) penicillin/streptomycin.
14. L929 conditioned medium (*see* **Note 1**).

15. Poly-L-lysine hydrobromide (PLL)-coated T75 cell culture flasks (*see* **Note 3**).
16. 0.05 % Trypsin/EDTA solution.
17. Hank's Balanced Salt Solution (HBSS).
18. Newborn mice at postnatal day 0–3.
19. 2 forceps with straight tips, tip size 0.05 mm × 0.01 mm, sterilized.
20. 2 forceps with curved tips, tip size 0.17 mm × 0.1 mm, sterilized.
21. Surgical scissor, 120 mm, sterilized.
22. Iris scissor, 90 mm, sterilized.
23. Micro scoop, round bottom, 140 mm, sterilized.
24. Sterile cell culture treated petri dishes (5 cm/10 cm).
25. Sterile 50 and 15 mL conical tubes.

2.2 Preparation of Monomeric, Oligomeric, and Fibrillar β-Amyloid (Aβ)

1. 1.5 mL Protein LoBind tubes (Eppendorf).
2. Fume hood.
3. Ice and ice buckets.
4. Lyophilized synthetic amyloid beta (Aβ) 1-42 peptide (Anaspec).
5. HFIP (1,1,1,3,3,3-Hexafluoro-2-propanol).
6. Speed vacuum centrifuge.
7. Sterile Dimethyl Sulfoxide (DMSO).
8. Water bath sonicator.
9. Phenol-red free DMEM.
10. Sterile distilled endotoxin-free water.
11. Parafilm.
12. 500 mM glycine–NaOH, pH 8.3.
13. 100 μM Thioflavin T in ultrapure water.
14. Black 384-well plate.
15. Fluorescence plate reader.

2.3 Stimulation of Microglia with Aβ for Inflammasome-Associated Assays (e.g., IL1β ELISA)

1. Cell culture medium: DMEM high glucose supplemented with L-glutamine, 10 % (v/v) FBS, 1 % (v/v) penicillin/streptomycin.
2. 96-well cell culture plate or appropriate cell culture dish for preferred detection method.
3. Ultrapure LPS.

3 Methods

3.1 Preparation and Cultivation of Primary Microglial Cells (See Note 2)

1. Sacrifice four newborn mice according to an approved animal protocol.
2. Decapitate mice by cutting neck with a surgical scissor.
3. Open both skull and skin following the midline with an iris scissor.
4. Remove the skull with curved forceps and transfer the brain with a micro scoop into a petri dish with pre-cooled HBSS under the binocular microscope.
5. Remove the cerebellum, split the hemispheres and remove meninges using forceps with straight tips.
6. Transfer brain hemispheres with the micro scoop into a 15 mL tube containing 5 mL HBSS (the volume should be sufficient to cover the hemispheres) and keep on ice until hemispheres of all four mice are collected in one tube.
7. Carefully aspirate HBSS and add 4 mL 0.05 % Trypsin-EDTA (1 mL per mouse brain).
8. Incubate in a water bath at 37 °C for 10 min.
9. Invert the tube several times during incubation.
10. Aspirate the Trypsin-EDTA solution and add 10 mL of pre-warmed cell culture medium.
11. Mechanically dissociate hemispheres by pipetting up and down with a sterile 10 mL pipette.
12. Split cell suspension between two PLL-coated T75 cell culture flasks (*see* **Note 3**), each containing 10 mL pre-warmed cell culture medium (3–4 hemispheres/flask) and incubate in cell culture incubator at 37 °C overnight.
13. The next day, remove cell culture medium, wash cells with pre-warmed sterile PBS to remove cell debris and add 10 mL of cell culture medium.
14. Incubate cells for 2–3 days in cell culture incubator at 37 °C until cell layer is confluent. The cell layer is mostly formed by astrocytes, while microglia and oligodendrocytes grow in a second layer above the astrocytes (*see* Fig. 1). Change medium and add 25 % (v/v) of L929 conditioned medium to induce microglial proliferation.
15. After 2–3 days, the microglial population is found loosely attached, on top of the astrocytic cell layer. In phase-contrast microscopy microglia appear round with a bright, shiny rim (*see* Fig. 1).
16. To harvest microglia, shake flasks horizontally (by hand), collect medium in a 50 mL tube and centrifuge cell suspension at $900 \times g$ for 10 min.

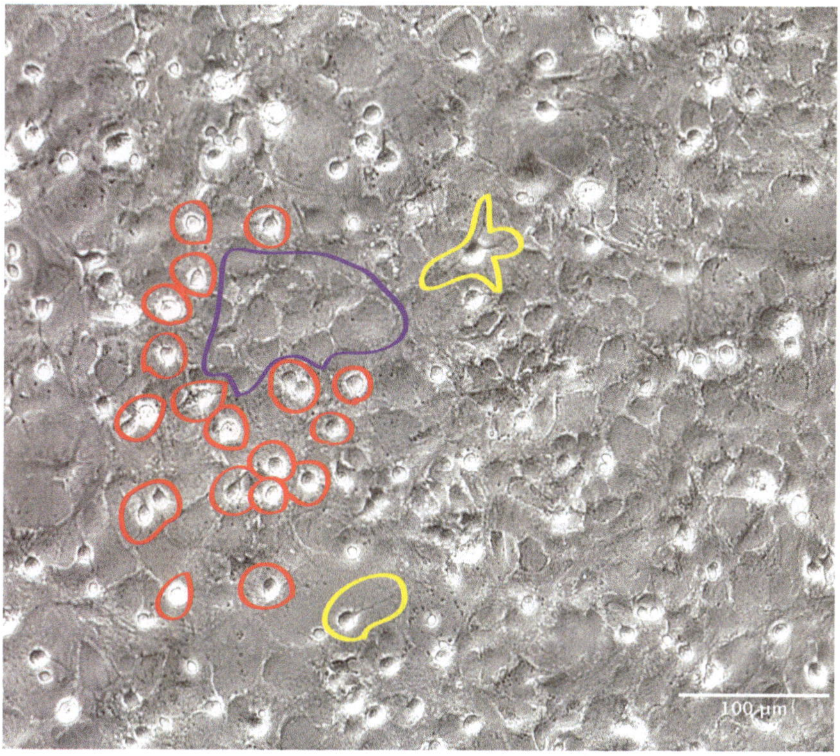

Fig. 1 Phase-contrast image of a mixed glial culture 7 days after preparation. Microglia and oligodendrocytes grow on a confluent layer of astrocytes. In phase-contrast microscopy microglia appear bright with a shiny rim. They display a round phenotype and are loosely attached, while oligodendrocytes are darker with long fine cell processes. As an example, cells demonstrating the distinct glial morphologies are marked: microglia (*red*), oligodendrocytes (*yellow*) and astrocytes (*purple*)

17. Re-suspend cell pellet in cell culture medium; count cells using a hemocytometer and use them for planned experiments. Depending on the quality of the glial culture a yield of $2–6 \times 10^6$ microglia/flask is expected (*see* **Note 4**). The purity of microglial cultures is usually above 95 % (*see* Fig. 2 and **Note 5**).

3.2 Preparation of Monomeric, Oligomeric, and Fibrillar β-Amyloid (Aβ)

3.2.1 Preparation of Aβ Monomers

1. Equilibrate lyophilized Aβ peptide to room temperature for 30 min (*see* **Note 6**).
2. Under the fume hood, dissolve lyophilized synthetic Aβ peptide in ice cold HFIP to a final concentration of 1 mM.
3. Incubate at room temperature for 2 h.
4. Aliquot 60 μL or appropriate volume into LoBind tubes.
5. Centrifuge LoBind tubes containing aliquots of Aβ/HFIP solution in a speed vacuum centrifuge at $800 \times g$ for 10–30 min at room temperature to remove HFIP (*see* **Note 7**). The clear peptide film becomes visible at the bottom of the tube.
6. Store tubes with peptide film at −80 °C for up to 6 months.

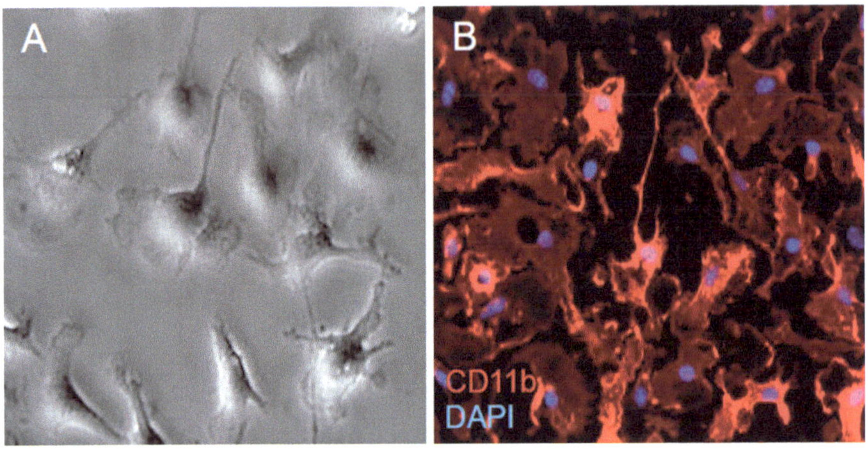

Fig. 2 Primary microglial cultures. (**a**) Primary microglial cells observed with (phase-contrast) live cell microscopy. (**b**) Paraformaldehyde-fixed microglial cells revealing a culture purity of approx. 98 % after the harvest. Cells were immunostained against microglial marker CD11b (*red*). Nuclei were stained with DAPI (*blue*)

3.2.2 Preparation of Oligomeric Aβ

1. Dissolve Aβ peptide film (from **step 6** of Subheading 3.2.1) in DMSO to a final concentration of 5 mM.
2. Sonicate in water bath sonicator for 5–10 min (*see* **Note 8**).
3. Add phenol-red free DMEM to reach a final concentration of 100 μM Aβ and vortex.
4. Incubate for 24 h at 4 °C.

3.2.3 Preparation of Fibrillar Aβ

1. Dissolve Aβ peptide film (from **step 6** of Subheading 3.2.1) in DMSO to a final concentration of 5 mM.
2. Sonicate in water bath sonicator for 5–10 min (*see* **Note 8**).
3. Add sterile distilled endotoxin-free water to reach a final concentration of 100 μM Aβ and vortex.
4. Seal tube with parafilm and incubate for 1 week at 37 °C.

3.2.4 Quality Control of Aβ Preparations Using a Thioflavin T Assay (See Note 9)

1. Vortex Aβ preparation (from **step 4** of Subheadings 3.2.2 or 3.2.3) and transfer 10 μL to 90 μL phenol-red free DMEM for a final concentration of 10 μM Aβ.
2. Pipette 80 μL of the Aβ-DMEM mixture, 10 μL 500 mM glycine–NaOH and 10 μL 100 μM Thioflavin T into one well of a black 384-well plate (*see* **Note 10**).
3. Measure fluorescence at 450/482 nm (excitation and emission maxima of Thioflavin T) in a fluorescence plate reader (*see* **Note 11**).

3.3 Stimulation of Primary Microglia with Aβ for IL-1β ELISA

1. After primary microglia are harvested from the astrocytic underlayer, seed 5×10^4 cells in a 96-well plate in 100 μL/well cell culture medium and incubate overnight in a 37 °C cell culture incubator (*see* **Note 12**).
2. The next day, remove medium and prime the cells with 100 ng/mL ultrapure LPS for 1–3 h in cell culture medium.

3. Add fibrillar or oligomeric Aβ at a final concentration of 10 μM and incubate for 6 h.
4. Collect supernatants and proceed to IL-β ELISA, as per the manufacturer's instructions.

4 Notes

1. The mouse cell line L929 (ATCC number: CCL-1) is used as a source of macrophage-colony stimulating factor (M-CSF). To obtain L929 cell-conditioned medium, L929 cells are seeded in T75 flasks and cultivated in DMEM medium with 10 % (v/v) FBS, 1 % (v/v) penicillin/streptomycin. The cells will grow and become confluent after a few days. The conditioned medium is collected after 7–14 days, centrifuged and filtered to remove cell debris and the supernatant is stored at -20 °C until use.
2. This is an example protocol for the preparation of microglia from four newborn mice. Scale material and solutions up or down according to the experimental requirements. Note that 3-4 hemispheres are needed per T75 cell culture flask and that the yield of microglia is relatively low (usually between approximately $2–6 \times 10^6$ cells per flask). It is therefore advised to prepare microglia from at least one mouse litter.
3. For PLL-coating of cell culture flasks prepare a PLL solution of 100 μg/mL in sterile water. Add 2–3 mL to the flask and make sure that the entire surface is covered. Incubate the flasks for at least 2 h at RT, aspirate the PLL solution and wash once with PBS.
4. After microglia have been harvested by shaking, the remaining mixed glial cultures can be re-used. To harvest microglia a second time, add fresh medium supplemented with 25 % (v/v) L929-conditioned medium after the first harvest and incubate cells 2–4 days at 37 °C to allow proliferation of remaining microglia.
5. The purity of the microglial culture can be assessed by flow cytometry using fluorophore-coupled CD11b antibodies to label microglia. Alternatively, microglia can be seeded onto glass coverslips, fixed with paraformaldehyde and immunostained with antibodies recognizing microglial markers CD11b or ionized calcium binding adaptor molecule 1 (Iba1).
6. Equilibration avoids condensation upon opening the vial [9].
7. Temperature inside the speed vacuum centrifuge should not exceed 25 °C [9].
8. Use slightly cooled water to prevent heating, but avoid freezing of DMSO during sonication.
9. Oligomer and fibril formation can additionally be tested by other approaches, such as transmission electron microscopy

(TEM) or gel electrophoresis and Western blotting (with an anti-Aβ 6E10 antibody) as previously described [10].

10. The final concentration of glycine–NaOH should be 50 mM, while the ratio between Thioflavin T and Aβ should be ≥1.
11. To monitor the aggregation state of Aβ take samples at specific time points and measure them immediately (within 5–10 min). Do not measure Aβ/Thioflavin T mixtures after longer time periods as Thioflavin T and glycine–NaOH may influence the aggregation of Aβ, thereby not resembling the Aβ aggregation state in the preparation tube. Thioflavin T binds to beta-sheet rich regions, which are more prevalent in Aβ fibrils than in monomeric/oligomeric Aβ. Therefore, fibrillar Aβ preparations should show about fourfold higher fluorescence intensities as compared to oligomeric preparations and eightfold higher as compared to negative controls (phenol-red free DMEM or water only).
12. Cell number and cell culture dish are chosen for an IL-1β ELISA. Adjust cell number and volume of cell culture medium according to your preferred assay (e.g., IL-1β or Caspase-1 Western Blot, FLICA assay)

References

1. Dahlgren KN, Manelli AM, Stine WB, Baker LK, Krafft G, LaDu MJ (2002) Oligomeric and fibrillar species of amyloid-beta peptides differentially affect neuronal viability. J Biol Chem 277(35):32046–32053. doi:10.1074/jbc.M201750200
2. Tomic JL, Pensalfini A, Head E, Glabe CG (2009) Soluble fibrillar oligomer levels are elevated in Alzheimer's disease brain and correlate with cognitive dysfunction. Neurobiol Dis 35(3):352–358. doi:10.1016/j.nbd.2009.05.024
3. Mandrekar S, Jiang Q, Lee CY, Koenigsknecht-Talboo J, Holtzman DM, Landreth GE (2009) Microglia mediate the clearance of soluble Abeta through fluid phase macropinocytosis. J Neurosci 29(13):4252–4262. doi:10.1523/JNEUROSCI.5572-08.2009, doi:29/13/4252 [pii]
4. Bolmont T, Haiss F, Eicke D, Radde R, Mathis CA, Klunk WE, Kohsaka S, Jucker M, Calhoun ME (2008) Dynamics of the microglial/amyloid interaction indicate a role in plaque maintenance. J Neurosci 28(16):4283–4292. doi:10.1523/JNEUROSCI.4814-07.2008, doi:28/16/4283 [pii]
5. Simard AR, Soulet D, Gowing G, Julien JP, Rivest S (2006) Bone marrow-derived microglia play a critical role in restricting senile plaque formation in Alzheimer's disease. Neuron 49(4):489–502. doi:10.1016/j.neuron.2006.01.022, doi:S0896-6273(06)00075-4 [pii]
6. Griffin WS, Stanley LC, Ling C, White L, MacLeod V, Perrot LJ, White CL 3rd, Araoz C (1989) Brain interleukin 1 and S-100 immunoreactivity are elevated in Down syndrome and Alzheimer disease. Proc Natl Acad Sci U S A 86(19):7611–7615
7. Blum-Degen D, Muller T, Kuhn W, Gerlach M, Przuntek H, Riederer P (1995) Interleukin-1 beta and interleukin-6 are elevated in the cerebrospinal fluid of Alzheimer's and de novo Parkinson's disease patients. Neurosci Lett 202(1–2):17–20
8. Halle A, Hornung V, Petzold GC, Stewart CR, Monks BG, Reinheckel T, Fitzgerald KA, Latz E, Moore KJ, Golenbock DT (2008) The NALP3 inflammasome is involved in the innate immune response to amyloid-beta. Nat Immunol 9(8):857–865
9. Fa M, Orozco IJ, Francis YI, Saeed F, Gong Y, Arancio O (2010) Preparation of oligomeric beta-amyloid 1-42 and induction of synaptic plasticity impairment on hippocampal slices. J Vis Exp (41), e1884, doi:10.3791/1884 (2010)
10. Jan A, Hartley DM, Lashuel H (2010) Preparation and characterization of toxic Abeta aggregates for structural and functional studies in Alzheimer's disease research. Nat Protoc 5(6):1186–1209

Chapter 2

Activating the NLRP3 Inflammasome Using the Amyloidogenic Peptide IAPP

Clara Westwell-Roper, Aisling Dunne, Man Lyang Kim, C. Bruce Verchere, and Seth L. Masters

Abstract

In addition to several other extracellular substances, phagocytosis of amyloid-forming peptides can perturb cellular homeostasis, leading to activation of the cytoplasmic innate immune receptor NLRP3. Once triggered, NLRP3 forms an inflammasome complex that ultimately cleaves pro-IL-1β and pro-IL-18 into their mature, secreted forms. Here we describe a protocol by which one type of amyloidogenic peptide, islet amyloid polypeptide (IAPP, otherwise known as amylin) can be prepared and used to stimulate myeloid cells in vitro to engage the NLRP3 inflammasome. Methods for measuring the ensuing inflammasome activation are also described. Although initially soluble, IAPP monomers rapidly aggregate in solution to form oligomers and subsequently insoluble amyloid fibrils. More work is required to examine how this transition influences inflammasome activation for different types of amyloid. The course of amyloid formation and corresponding inflammatory capacity of these pre-fibrillar species following uptake also requires further examination, and we hope that our protocols are useful in these endeavors. While these protocols are restricted to examination of synthetic IAPP, isolation of IAPP aggregates from human and transgenic mouse pancreas will be required to definitively determine the proinflammatory effects of endogenous IAPP oligomers and fibrils.

Key words Inflammasome, Amyloid, Innate immunology, NLR, IAPP

1 Introduction

Activation of innate immune cells by several different types of amyloid-forming peptides has been reported for a number of years and may be mediated, in part, by cell surface Toll-like receptors [1, 2]. Recently, it has emerged that the cytoplasmic innate immune receptor, NLRP3, operates in myeloid cells to detect the danger presented by these uniquely folded proteins [3–7]. Although the exact mechanism by which amyloid triggers NLRP3 remains unknown, the downstream process of apoptosis-associated speck-like protein containing a CARD (ASC) oligomerization, autocatalytic activation of caspase-1, and subsequent cleavage of IL-1β and IL-18, in a complex called the inflammasome, has been well

Christine M. De Nardo and Eicke Latz (eds.), *The Inflammasome: Methods and Protocols*, Methods in Molecular Biology, vol. 1040, DOI 10.1007/978-1-62703-523-1_2, © Springer Science+Business Media New York 2013

described [8]. Protocols to examine activation at several points in this pathway are described herein, namely, (a) measurement of caspase-1 activity using a fluorescent caspase-1 substrate and (b) Western blotting and ELISA measurement of mature, secreted IL-1β and IL-18, respectively. Furthermore, we describe preparation of the materials required (putative oligomers of IAPP capable of forming amyloid), and a protocol for detecting this amyloid when it is phagocytosed by macrophages.

IAPP is a 37 amino acid peptide that is co-secreted with insulin by beta cells [9, 10]. Insoluble amyloid deposits comprised primarily of IAPP fibrils are found in more than 90 % of patients with type 2 diabetes at autopsy [11, 12]. Amyloid fibrils are defined by a cross-beta sheet quaternary structure with a characteristic X-ray diffraction pattern. Several lines of evidence suggest that soluble oligomers of IAPP rather than mature fibrils may be the major cytotoxic species and also an important stimulus for islet inflammation, although both fibrils and oligomers likely contribute to beta cell dysfunction in vivo [13]. We have recently provided evidence to suggest that IAPP acts as a trigger for macrophage recruitment and secretion of proinflammatory cytokines that are toxic to beta cells, suggesting that islet amyloid may be a major stimulus for islet inflammation observed in patients with type 2 diabetes [6, 7]. Oligomers of human IAPP share a common structure with numerous other disease-causing amyloids [14]. The inflammatory capacity of these amyloids and other protein aggregates is currently an area of intensive research, with significant clinical implications for diseases such as Alzheimer's disease, age-related macular degeneration, Huntington's disease, Parkinson's disease, and others [8]. Despite recent advances in theoretical and empirical modeling of amyloid formation and toxicity, further work is required to determine the biochemical basis for the pro-inflammatory propensity of amyloidogenic peptides both in vitro and in vivo.

In total we outline four methods. The first uses fluorescence microscopy to detect IAPP amyloid when present in macrophages, and could be modified to detect different species of amyloid in other cell types. This could also be examined contemporaneously with measurement of inflammasome activation such as ASC "speck" formation. Secondly, measurement of caspase-1 activation by IAPP is described using the FAM-FLICA Caspase Detection Kit. This assay is based on the irreversible binding of a cell permeable fluorescein labeled inhibitor to the active enzyme, which can be detected by fluorescence microscopy; however, we describe its application in flow cytometry. Thirdly, we briefly describe Western blot analysis of cleaved IL-1β after IAPP activation. Finally, we outline a protocol to measure secreted IL-18 by ELISA that is more affordable than commercially available ready-made kits. It is important that for all experiments, cytotoxicity is measured by propidium iodide, LDH or another method in parallel. This will ensure that the effect observed occurs at a time and concentration of IAPP that does not cause significant levels of cell death.

2 Materials

2.1 Preparation of IAPP Oligomers

1. Islet amyloid polypeptide (IAPP) (*see* **Note 1**).
2. Rodent IAPP (MW 3920).
3. 1,1,1,3,3,3-hexafluoro-2-propanol (HFIP) (Sigma).
4. 0.1 M acetic acid.
5. Sigmacote (Sigma).
6. Thioflavin T powder (Sigma): Dissolve in distilled water (dH_2O) to prepare a 100 μM (10×) stock solution prior to each assay. Pass the solution through a 0.2 μm filter to eliminate small aggregates (*see* **Note 2**). The stock solution can be stored in the dark at 4 °C for 1 month.
7. 0.2 μm filter.
8. Sterile 1.5 mL tubes.
9. Desiccator at −20 °C.
10. Nitrogen gas or lyophilizer.
11. 96-well plate for fluorescence measurement (clear bottom).
12. Fluorescence plate reader.

2.2 Detection of IAPP Amyloid in Macrophages by Fluorescent Microscopy

1. Phosphate buffered saline (PBS).
2. 4 % Paraformaldehyde.
3. 0.2 % Triton X-100.
4. 70 % Ethanol.
5. Distilled water (dH_2O).
6. Thioflavin S powder (Sigma) (*see* **Note 3**).
7. Bone marrow derived macrophages (BMDMs).
8. Microscope slides or glass-bottom plates for fluorescence microscopy.
9. Confocal or fluorescence microscope.

2.3 Analysis of Active Caspase-1 in Macrophages Using a Cleavable Fluorescent Substrate

1. BMDMs.
2. PBS, pH 7.4.
3. Dimethyl sulfoxide (DMSO).
4. 24-well plate.
5. Cell culture incubator at 37 °C under 5 % CO_2.
6. Cell scraper.

7. FACS tubes.
8. FAM-FLICA Caspase Detection Kit (ImmunoChemistry Technologies).
9. Flow cytometer with an argon ion laser.

2.4 Western Blotting for Cleaved, Secreted IL-1β

1. Ultrapure LPS from *E. coli* (Invivogen).
2. Methanol.
3. Chloroform.
4. 2× SDS loading buffer: 4 % SDS, 125 mM Tris–HCl, pH 6.8, 20 % glycerol, 0.02 % (w/v) bromophenol blue, 200 mM dithiothreitol.
5. 12 % or gradient SDS-polyacrylamide gel.
6. PVDF or nitrocellulose membrane.
7. Monoclonal rat anti-mouse IL-1β antibody (R&D Systems, Clone 166926).
8. Heat block for 50 and 95 °C incubations.

2.5 ELISA Analysis of Mouse IL-18

1. 96-well NUNC Maxisorp ELISA plate.
2. Wash buffer: PBS + 0.05 % Tween 20 (PBS-T).
3. Block: 1 % BSA + PBS-T.
4. Rat anti-mouse IL-18 capture antibody (clone 74).
5. Biotinylated rat anti-mouse IL-18 detection antibody (clone 93-10C).
6. Murine recombinant IL-18 cytokine standard (R&D systems, B004-5).
7. HRP-streptavidin.
8. TMB substrate solution.
9. Stop solution: 1 M H_2SO_4 or alternative.

3 Methods

3.1 Preparation of IAPP Oligomers (See Notes 1 and 4)

1. Dissolve 0.5 mg of lyophilized human IAPP (MW 3,903 Da) at a concentration of 500 μM by adding 256 μL of HFIP. Use rodent IAPP as a non-amyloidogenic control. Allow the peptide to dissolve in HFIP for several hours or overnight, and then aliquot as desired (e.g., 50 μL per tube in 1.5 mL tubes).
2. Remove the HFIP using nitrogen gas. To do this, gently blow the inert gas into the tube (*see* **Note 5**).
3. Resuspend an aliquot in 5 μL of 0.1 M acetic acid (*see* **Note 6**).

4. This can then be diluted in H_2O, PBS, or media to the same volume as before the HFIP was removed, to return a 500 μM concentration, which we typically use as a 50× stock (*see* **Note** 7).
5. Measure IAPP aggregation using Thioflavin T fluorescence. Dissolve IAPP at the desired concentration in media under conditions used in parallel tissue culture experiments. Add Thioflavin T at a final concentration of 10 μM for a final volume of 50–100 μL/well in a 96-well plate (black with clear bottom) (*see* **Note 8**).
6. Incubate the plate at 37 °C and monitor amyloid formation by fluorescence for up to 24 h at 450/482 nm (Thioflavin T excitation/emission peaks) (*see* **Notes 9** and **10**). Ensure the plate is well sealed to prevent evaporation.

3.2 Detection of IAPP Amyloid in Macrophages by Fluorescence Microscopy

1. Seed cells of interest such as bone marrow derived macrophages (BMDMs) on multi-well glass microscope chamber slides or a 96-well plate for fluorescence microscopy (10^5 cells/well).
2. Allow cells to adhere overnight at 37 °C.
3. Treat cells with human IAPP (use rat IAPP as a non-amyloidogenic control) from 0 to 10 μM over 0–24 h to assess peptide uptake and aggregation (*see* **Note 11**).
4. Wash cells with PBS and fix with 4 % paraformaldehyde for 15 min at room temperature (RT). Use sufficient paraformaldehyde to cover the cells completely (e.g., 50 μL/well for a 96-well plate or 500 μL for an 8-well chamber slide).
5. Wash cells with PBS and then store at 4 °C until ready for staining.
6. Permeabilize cells by adding sufficient 0.2 % Triton X-100 to immerse them entirely for 5 min at RT.
7. Wash 3× in PBS for 5 min each wash.
8. Incubate cells with 0.01 % Thioflavin S for 10 min at RT. Use enough Thioflavin S to fully immerse fixed cells.
9. Wash stained cells 3× for 5 min each with 70 % ethanol followed by dH_2O (*see* **Note 12**).
10. Mount cells as desired and examine by confocal or fluorescence microscopy (*see* **Note 13**).

3.3 Analysis of Active Caspase-1 in Macrophages Using a Cleavable Fluorescent Substrate

1. Seed BMDMs in a 24-well plate (500 μL/well) at a density of 1×10^6 per mL, the day before carrying out this assay.
2. Treat cells with 10 μL/well of a 500 μM IAPP stock in order to give a final concentration of 10 μM and incubate for 1 h at 37 °C (*see* **Note 14**).
3. Detach cells from the base of the wells using a cell scraper and transfer to FACS tubes.

4. Reconstitute the FLICA reagent with 50 μL of DMSO to generate a 150× stock solution (*see* **Note 15**). Dilute the required amount of 150× stock solution 1:5 in PBS immediately prior to use; this will form a 30× stock solution.
5. Add 16 μL/well of 30× FLICA reagent to IAPP treated and untreated cells and incubate for 1 h at 37 °C under 5 % CO_2 (Unlabeled IAPP treated and untreated cells should be included as additional controls).
6. Centrifuge the plate at <400 × *g* for 5 min at RT.
7. Prepare the required amount of wash buffer by diluting the 10× wash buffer stock solution from the FLICA kit 1:10 with dH_2O.
8. Wash cells with 2 mL of 1× wash buffer and centrifuge at <400 × *g* for 5 min at RT. Repeat this step twice more.
9. Discard the supernatant and resuspend cells in 300 μL of 1× wash buffer. Run cells on a flow cytometer with an argon ion laser at 488 nm and detect in the FL1 channel (515–535 nm) (*see* **Note 16**).

3.4 Western Blotting for Cleaved, Secreted IL-1β

1. Seed BMDMs in a 24-well plate (500 μL/well) at a density of 1×10^6 per mL, in a low concentration of FCS, the day before carrying out this assay (*see* **Note 17**).
2. Prime cells with 100 ng/mL ultrapure LPS for 3 h at 37 °C.
3. Stimulate cells with 10 μM IAPP overnight.
4. Collect the supernatant and mix with 500 μL methanol and 125 μL chloroform to precipitate all proteinaceous material.
5. Centrifuge the sample at 16,000 × *g* for 5 min and then remove the top layer, leaving the precipitate at the interface and a lower liquid layer.
6. Add 500 μL of methanol and resuspend.
7. Centrifuge the sample for 5 min at 16,000 × *g* and then aspirate the supernatant and dry the pellet at 50 °C for 5 min.
8. Resuspend the pellet in 50–100 μL 2× SDS loading buffer, and mix and heat at 95 °C until the pellet is dissolved.
9. Run the sample on a low percentage or gradient SDS-polyacrylamide gel, then transfer to PVDF or nitrocellulose, and blot using an antibody such as rat anti-mouse IL-1β.

3.5 ELISA Analysis of Mouse IL-18

1. Coat each well of a 96-well NUNC Maxisorp ELISA plate with 100 μL IL-18 capture antibody diluted 1/1,000 in PBS (final concentration 1 μg/mL), overnight at 4 °C (*see* **Note 18**).
2. Wash the plate 2× with PBS-T (approximately 200 μL, can be filled by immersion).

3. Block the plate with 200 μL of 1 % BSA+PBS-T per well, for 2 h at RT.
4. Wash the plate as in **step 2** of Subheading 3.5.
5. Add samples and recombinant mouse IL-18 standard made up in 1 % BSA+PBS-T, then incubate the plate for 2 h at RT or overnight at 4 °C (*see* **Note 19**).
6. Wash the plate 5× with PBS-T then add 100 μL IL-18 detection antibody diluted 1/10,000 in 1 % BSA+PBS-T and incubate for 1 h at RT (*see* **Note 18**).
7. Wash the plate 5× with PBS-T, then add HRP-streptavidin and incubate for 30 min at RT.
8. Wash the plate 5× with PBS-T then add 100 μL TMB substrate solution per well and incubate for 5–10 min.
9. Add 50 μL stop solution per well and read absorbance at 450 nm, subtracting background reading at 620 nm.

4 Notes

1. Milligram quantities of IAPP peptide can be synthesized by traditional chemistries. Alternatively, these can be purchased from commercial sources, such as Bachem or Merck. When working with IAPP, and other amyloid species, it may be important to use glassware, tips, and other equipment that are treated to minimize adherence. In this protocol, when resuspending IAPP, using tips that are treated with Sigmacote promotes recovery.
2. Thioflavin T is a benzothiazole salt that undergoes a red shift of its emission spectrum upon binding to beta sheet-rich structures. The increase in emission at 482 nm can be used to monitor amyloid formation [15]. Fluorescence correlates with fibril formation as determined by electron microscopy, although Thioflavin T may also bind some pre-fibrillar IAPP aggregates [16].
3. Thioflavin S, similar to Thioflavin T, binds to beta sheet-rich structures and is particularly well suited for detecting intracellular IAPP amyloid by fluorescence microscopy. Thioflavin S powder can be purchased from Sigma and stock solutions prepared at a concentration of 0.5 % in dH_2O followed by passage through a 0.2 μm filter.
4. Appropriate preparation of IAPP is critical to obtain species that are capable of activating the inflammasome. In our hands, freshly prepared soluble IAPP must be quickly added to macrophages to have this effect, while the insoluble fibrils that form over time do not. However, this may not necessarily be the case for other types of amyloid, and further work is required

to characterize the nature of the IAPP aggregate responsible for NLRP3 activation.

5. When removing the HFIP using the nitrogen gas, take care not to use excessive force, as this would cause the liquid to spill, or evaporate over a large surface area. Alternatively, the sample can be lyophilized until the peptide becomes a dry, fluffy white powder. When completely dry, the peptide can be stored at −20 °C in a desiccator for extended periods of time.
6. Resuspending the aliquot in 0.1 M acetic acid improves the solubility of IAPP.
7. Resuspending this aliquot in H_2O, the IAPP will persist as oligomers longer than if resuspended in a buffer such as PBS or media, which favors fibril formation [6]. We see maximum NLRP3 inflammasome activation when this material is quickly added to macrophages or dendritic cells after being reconstituted in H_2O.
8. Thioflavin T can be used at a final concentration of 10 μM in media, PBS, H_2O, or Tris–HCl, pH 7.4. Parameters such as IAPP concentration, temperature, agitation, dissolved solutes, pH, and the presence of cell membranes all affect aggregation kinetics.
9. The kinetics of aggregation should be determined in parallel with each culture experiment to ensure consistency among experiments in relation to time points for cell treatment and lysis. Amyloid formation typically occurs within several hours at 37 °C.
10. Cells may be treated with IAPP at different times following peptide dissolution to evaluate the effects of early and late aggregates on inflammasome activation. However, more precise characterization of aggregation state may be achieved using methods such as atomic force microscopy, SDS-PAGE, and binding to monoclonal anti-monomer/oligomer antibodies. Despite recent experimental and theoretical models of IAPP aggregation, there remains considerable uncertainty regarding the details of the steps of IAPP fibril formation.
11. We typically grow BMDMs for 7 days in DMEM + 10 % FCS + 10–20 % L-cell supernatant, on non-tissue-culture-treated 10 cm dishes. The cells can then be recovered by removing the medium, adding 10 mL cold PBS and incubating for 10 min at 4 °C, followed by scraping. These cells can then be seeded on tissue-culture-treated plates for use the next day.
12. If other stains will be used (e.g., lysosomal markers), these should be carried out prior to the Thioflavin S staining.
13. While uptake of IAPP fibrils is thought to occur by receptor-mediated phagocytosis, macropinocytosis is responsible for

clearance of soluble amyloid-β. It is unclear whether this is also the case for oligomeric forms of IAPP. Detection of prefibrillar IAPP in cultured cells may be achieved using commercially available antibodies that bind to IAPP monomers and/or oligomers.

14. Priming these cells with LPS does not seem to increase caspase-1 activity in this assay.
15. The 150× stock FLICA solution can be stored in the dark at −20 °C for up to 6 months. During this time it can be thawed and used twice.
16. If cells will not to be analyzed immediately: Add 30 μL of fixative (2 % paraformaldehyde in PBS, pH 7.4), incubate in the dark for 15 min at room temperature and store fixed cells at 2–8 °C for up to 24 h.
17. For this protocol, cells are grown in a low concentration of FCS as excess protein can overload an SDS-polyacrylamide gel. As we typically use BMDMs that are supplemented in 10–20 % L-cell supernatant containing FCS, this represents a final concentration of 1–2 % FCS which is suitable. A minimum of 500 μL supernatant is desirable.
18. The capture and detection antibodies should both be titrated for robust and sensitive detection of IL-18 in each laboratory.
19. We normally use a concentration of 2 ng/mL IL-18 for the top standard and dilute this twofold sequentially across 8–12 wells.

Acknowledgments

The authors would like to thank Ben A. Croker for initially optimizing the IL-18 ELISA protocol. S.L.M. is supported by a fellowship from the National Health and Medical Research Council of Australia (516783). C.W.R. is supported by a Vanier Canada Graduate Scholarship and C.B.V. by grants from the Canadian Institutes of Health Research (MOP-14862) and 20R66631 and the Canadian Diabetes Association (OG-3-11-3413-CV).

References

1. Tukel C, Wilson RP, Nishimori JH, Pezeshki M, Chromy BA, Baumler AJ (2009) Responses to amyloids of microbial and host origin are mediated through toll-like receptor 2. Cell Host Microbe 6(1):45–53. doi:10.1016/j.chom.2009.05.020, doi:S1931-3128(09)00213-3 [pii]
2. Stewart CR, Stuart LM, Wilkinson K, van Gils JM, Deng J, Halle A, Rayner KJ, Boyer L, Zhong R, Frazier WA, Lacy-Hulbert A, El Khoury J, Golenbock DT, Moore KJ (2010) CD36 ligands promote sterile inflammation through assembly of a Toll-like receptor 4 and 6 heterodimer. Nat Immunol 11(2):155–161. doi:10.1038/ni.1836, doi:ni.1836 [pii]
3. Ather JL, Ckless K, Martin R, Foley KL, Suratt BT, Boyson JE, Fitzgerald KA, Flavell RA, Eisenbarth SC, Poynter ME (2011) Serum

amyloid A activates the NLRP3 inflammasome and promotes Th17 allergic asthma in mice. J Immunol 187(1):64–73. doi:10.4049/jimmunol.1100500, doi:jimmunol.1100500 [pii]
4. Halle A, Hornung V, Petzold GC, Stewart CR, Monks BG, Reinheckel T, Fitzgerald KA, Latz E, Moore KJ, Golenbock DT (2008) The NALP3 inflammasome is involved in the innate immune response to amyloid-beta. Nat Immunol 9(8):857–865. doi:10.1038/ni.1636, doi:ni.1636 [pii]
5. Niemi K, Teirila L, Lappalainen J, Rajamaki K, Baumann MH, Oorni K, Wolff H, Kovanen PT, Matikainen S, Eklund KK (2011) Serum amyloid A activates the NLRP3 inflammasome via P2X7 receptor and a cathepsin B-sensitive pathway. J Immunol 186(11):6119–6128. doi: 10.4049/jimmunol.1002843, doi:jimmunol.1002843 [pii]
6. Masters SL, Dunne A, Subramanian SL, Hull RL, Tannahill GM, Sharp FA, Becker C, Franchi L, Yoshihara E, Chen Z, Mullooly N, Mielke LA, Harris J, Coll RC, Mills KH, Mok KH, Newsholme P, Nunez G, Yodoi J, Kahn SE, Lavelle EC, O'Neill LA (2010) Activation of the NLRP3 inflammasome by islet amyloid polypeptide provides a mechanism for enhanced IL-1beta in type 2 diabetes. Nat Immunol 11(10):897–904. doi:10.1038/ni.1935, doi:ni.1935 [pii]
7. Westwell-Roper C, Dai DL, Soukhatcheva G, Potter KJ, van Rooijen N, Ehses JA, Verchere CB (2011) IL-1 blockade attenuates islet amyloid polypeptide-induced proinflammatory cytokine release and pancreatic islet graft dysfunction. J Immunol. doi:10.4049/jimmunol.1002854
8. Masters SL, O'Neill LA (2011) Disease-associated amyloid and misfolded protein aggregates activate the inflammasome. Trends Mol Med 17(5):276–282. doi:10.1016/j.molmed.2011.01.005
9. Kahn SE, D'Alessio DA, Schwartz MW, Fujimoto WY, Ensinck JW, Taborsky GJ Jr, Porte D Jr (1990) Evidence of cosecretion of islet amyloid polypeptide and insulin by beta-cells. Diabetes 39(5):634–638
10. Westermark P, Wernstedt C, Wilander E, Sletten K (1986) A novel peptide in the calcitonin gene related peptide family as an amyloid fibril protein in the endocrine pancreas. Biochem Biophys Res Commun 140(3):827–831
11. Westermark P (1972) Quantitative studies on amyloid in the islets of Langerhans. Ups J Med Sci 77(2):91–94
12. Bell ET (1959) Hyalinization of the islets of Langerhans in nondiabetic individuals. Am J Pathol 35(4):801–805
13. Zraika S, Hull RL, Verchere CB, Clark A, Potter KJ, Fraser PE, Raleigh DP, Kahn SE (2010) Toxic oligomers and islet beta cell death: guilty by association or convicted by circumstantial evidence? Diabetologia 53(6):1046–1056. doi:10.1007/s00125-010-1671-6
14. Kayed R, Head E, Thompson JL, McIntire TM, Milton SC, Cotman CW, Glabe CG (2003) Common structure of soluble amyloid oligomers implies common mechanism of pathogenesis. Science 300 (5618):486–489. doi:10.1126/science.1079469300/5618/486 [pii]
15. Park K, Verchere CB (2001) Identification of a heparin binding domain in the N-terminal cleavage site of pro-islet amyloid polypeptide. Implications for islet amyloid formation. J Biol Chem 276(20):16611–16616. doi:10.1074/jbc.M008423200 M008423200 [pii]
16. Kudva YC, Mueske C, Butler PC, Eberhardt NL (1998) A novel assay in vitro of human islet amyloid polypeptide amyloidogenesis and effects of insulin secretory vesicle peptides on amyloid formation. Biochem J 331(Pt 3):809–813

Chapter 3

Assessment and Quantification of Crystal-Induced Lysosomal Damage

Peter Duewell and Eicke Latz

Abstract

Lysosomes are organelles that degrade endocytosed, phagocytosed, or autophagocytosed materials. Lysosomal degradation of engulfed material is a highly controlled mechanism, which is vital in the control of infection, recycling of cellular organelles, and the breakdown of larger material. Following ingestion, lysosomes acidify and mature, leading to the activation of a range of proteolytic enzymes. Lysosomes are considered stable organelles that separate their highly hydrolytic enzymes from the cytosol to prevent digestion of self-molecules and host cell damage. Some substances, mostly of a particulate or aggregated nature, are able to damage lysosomes. Lysosomal damage and rupture, with subsequent release of lysosomal content into the cytosol can have dramatic consequences for the cell, such as the induction of cell death or activation of the NLRP3 inflammasome. In this chapter, we provide methods for the induction, assessment, and quantification of lysosomal damage by flow cytometry and confocal microscopy.

Key words Lysosomal damage, Lysosomal translocation, Confocal microscopy, Flow cytometry, FACS, NLRP3, Inflammasome, Caspase-1, Inflammation

1 Introduction

All cells require lysosomes for the defense against invading pathogens and to degrade endogenous or exogenous materials. Furthermore, in antigen-presenting cells, such as dendritic cells and macrophages, lysosomes have important immunological functions, as they are required for antigen processing and presentation. Lysosomes contain a multitude of digestive enzymes such as lipases, nucleases and proteases that function in waste disposal of endogenous, as well as exogenous material and cell debris. These enzymes only gain their lytic function in an acidic milieu that is facilitated by ATP-dependent proton pumps located on the lysosomal membrane. Lysosomal degradation is a controlled cellular process. Highly glycosylated proteins at the inner lysosomal membrane and the neutral pH of the cytosol regularly minimize enzymatic self-digestion. Under certain conditions, the integrity of

Christine M. De Nardo and Eicke Latz (eds.), *The Inflammasome: Methods and Protocols*, Methods in Molecular Biology, vol. 1040, DOI 10.1007/978-1-62703-523-1_3, © Springer Science+Business Media New York 2013

lysosomes can be compromised and the ensuing leakage of lysosomal content into the cytosol can cause harm to the cell. Since lysosomal destabilization can jeopardize the survival of cells, lysosomes were also nicknamed "suicide bags". Chemical compounds, such as L-leucyl L-leucine methyl ester, are taken up by lysosomes and form membranolytic complexes via dipeptidyl peptidase I and induce lysosomal damage, leading to the release of lysosomal content into the cytosol [1]. Recently, it has been shown that disease-related particulate substances are also able to induce lysosomal maturation and rupture, and this process was implicated in cellular demise as well as activation of the NLRP3 inflammasome [2–4]. It has been shown that leakage of acidic and hydrolytic lysosomal content into the cytosol can set off an inflammatory response. Fluorescence methods, such as flow cytometry and combined confocal laser reflection and fluorescence imaging, allow for studying the uptake of lysosomal damaging substances, the induction of lysosomal damage, and the translocation of lysosomal content into the cytosol. Furthermore, lysosomal damage can be mimicked and measured in vitro using chemical substances that can damage lysosomes.

In this chapter, we describe how to induce lysosomal damage by incubating cells with L-leucyl-L-leucine methyl ester (Leu-Leu-OMe) or crystalline particles, such as silica crystals. We outline how to visualize and track the process of lysosomal maturation following damage using fluorescent dyes. Furthermore, we exemplify how this process can be interrupted using specific inhibitors of lysosomal function.

2 Materials

2.1 Cell Culture and Media

1. Immortalized macrophages (iMØ) derived from C57BL/6 mice (*see* **Note 1**) [2].
2. Complete medium: Dulbecco's modified Eagle Medium (DMEM), supplemented with 10 % fetal bovine serum (FBS), 1 % L-glutamine and 10 μg/mL ciprofloxacin.
3. Phosphate buffered saline (PBS).
4. 1× Trypsin/EDTA.

2.2 Reagents

Use all reagents at room temperature. Stock solutions are further diluted as indicated in the Subheadings 3 and 4.

1. 50 mg/mL Silica (MIN-U-SIL-15, US Silica, Frederick, USA), in PBS.
2. 1 M L-leucyl L-leucine methyl ester (Leu-Leu-OMe) in H_2O (*see* **Note 2**).
3. 160 μM bafilomycin A1 in DMSO (*see* **Note 3**).

Table 1
Excitation and emission of selected dyes with the appropriate laser used at the confocal microscope

Dye conjugate	Excitation (nm)	Emission (nm)	Laser
Acridine orange	460	600–650	Ar (488 nm)
Alexa-488-Dextran	495	519	Ar (488 nm)
DQ-Ovalbumin	505	515	Ar (488 nm)
LysoTracker® Green	443	505	Ar (488 nm)
Alexa-647-CtxB	650	668	HeNe (633 nm)

Ar argon laser with 488 nm wavelength, *HeNe* helium/neon laser with 633 nm wavelength (*see* **Note 23**)

Table 2
Fluorescent dye-labeled agents with their functions in cell staining

Dye conjugate	Staining function in cells
Acridine orange	Acridine orange is a cationic molecule and accumulates in compartments of low pH and forms dimers that appear red. It also binds RNA and DNA in its monomeric state resulting in green fluorescence
Alexa-488-Dextran	Dye-labeled Dextran is a highly water-soluble polysaccharide that likely accumulates in vesicles. Due to its uncommon a-1–6-linkage, Dextran is mostly inert for glucosidases and allows long-term acquisition
DQ-Ovalbumin	DQ-Ovalbumin is a quenched conjugate consisting of a pH-insensitive dye and Ovalbumin. Following endocytosis the molecule is proteolytically cleaved and the accumulation of the dye results in a bright fluorescence
LysoTracker® Green	LysoTracker is a fluorescent acidotropic dye coupled to a basic amine. Less protonated at low pH, these molecules likely permeate and accumulate in vesicles. Due to the low pH in lysosomes, LysoTracker becomes protonated and fluorescent
Alexa-647-CtxB	Fluorochrome-conjugated cholera toxin subunit B is the recombinant nontoxic subunit B that is important for the protein complex to bind on cell surface and allows membrane staining

4. 1 mg/mL Alexa-Fluor 647-conjugated cholera toxin subunit B (CtxB) in H_2O.
5. 10 mg/mL acridine orange (AO), in H_2O (*see* Tables 1 and 2).
6. 5 mg/mL DQ-Ovalbumin (DQ-OVA), in H_2O (*see* Tables 1 and 2).
7. 5 mg/mL Alexa-Fluor 488-conjugated dextran (488-Dex), MW 10.000, in H_2O (*see* Tables 1 and 2).
8. 1 mM LysoTracker® *Green* (LTG) (Invitrogen), in DMSO (*see* Tables 1 and 2).

2.3 Equipment

1. 12-well flat-bottom tissue culture plates.
2. Glass bottom dishes for confocal microscopy.
3. Confocal laser-scanning microscope.
4. Flow cytometer.
5. Cell culture incubator at 37 °C and 5 % CO_2.
6. Serological pipettes.
7. Micropipette tips.
8. Empty tip box for "humidity chamber".
9. 1.5 mL and 15 mL tubes with lid.
10. FACS tubes.
11. Computer with FlowJo (Treestar) and Adobe Photoshop (Adobe) software.

3 Methods

3.1 Quantification of Lysosomal Loss of Fluorescence by FACS

1. Seed 0.1–1 × 10^5 iMØ cells/mL into a 12-well plate (1 mL/well).
2. Incubate cells for at least 2 h at 37 °C to allow for adherence (*see* **Note 4**).
3. Replace the medium with 450 μL of fresh DMEM/well (*see* **Notes 5** and **6**).
4. Add 50 μL of a 50 μg/mL AO solution/well to obtain a final concentration of 5 μg/mL (*see* **Notes 7** and **8**).
5. Incubate the plate for 15 min at 37 °C.
6. Wash the cells 3× with 500 μL PBS/wash (*see* **Note 8**).
7. Remove the PBS and replace it with 450 μL DMEM/well.
8. Add 50 μL of a 10× concentration of Silica (5 mg/mL) or 50 μL of a 10× concentration of Leu-Leu-OMe (1 μM) per well (*see* **Note 9**).
9. Incubate the plate for 6 h at 37 °C (*see* **Note 10**).
10. Remove the medium, add 100 μL 1× Trypsin/EDTA/well and incubate the plate for 3 min at 37 °C (*see* **Note 11**).
11. Stop the trypsinization by adding 400 μL complete medium/well and transfer the cell suspension into FACS collection tubes.
12. Centrifuge tubes for 5 min at 400 × *g* and resuspend the pellet in 150 μL PBS.
13. Enable the PE-filter channel on your FACS machine (*see* **Note 12**).
14. Begin by acquiring the untreated control sample.

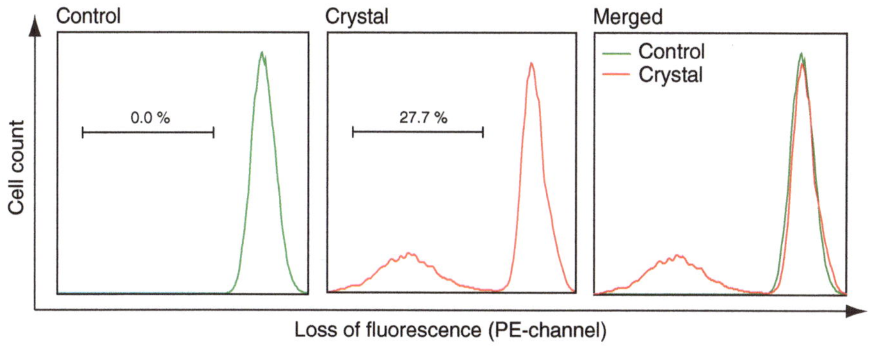

Fig. 1 Lysosomal loss of fluorescence. Cells were labeled with acridine orange prior to stimulation with cholesterol crystals. Loss of fluorescence was acquired with FACS analysis as fluorescence loss in the PE-channel

15. Exclude cell duplets by setting a gate on the main population obtained from a FSC-A vs. FSC-H dotplot (*see* **Note 13**).
16. Compare the PE-channel on the *x*-axis with the cell count (histogram).
17. Adjust the PE-channel voltage to obtain a peak at a fluorescence intensity of about 10^3–10^4 (*see* **Note 14**).
18. Acquire all samples.
19. Insert a gate starting from the *y* axis to the beginning of thc AO-peak and apply to all samples (*see* Fig. 1).
20. Determine the percentage of cells that have intermediate fluorescence in the PE channel using FlowJo software.

3.2 Assessing Lysosomal Rupture and Translocation by Confocal Microscopy

1. Seed 1×10^5 iMØ cells/mL into glass bottom dishes (500 μL/dish) (*see* **Note 15**).
2. Follow **steps 2–3** of Subheading 3.1. Optional: Inhibitors such as bafilomycin A1 could be used at this point (*see* **Notes 5** and **6**).
3. Label cells with 5 μg/mL AO, 50 μg/mL DQ-OVA, 20 μg/mL Alexa-488 dextran or 50 nM LysoTracker® green labeling medium (*see* **Note 16**, Tables 1 and 2).
4. Remove the supernatant and add 500 μL of the prepared labeling-medium.
5. Incubate the dish for 15 min (AO) or 1 h (DQ-OVA, 488-Dex, LysoTracker Green).
6. Follow **steps 7–10** of Subheading 3.1.
7. Prepare a "humidity chamber" by placing a moist paper towel in the bottom of an empty tip box (*see* **Note 17** and Fig. 2).
8. Remove the medium and add 100 μL of 10 μg/mL CtxB-containing medium directly onto the glass-bottom area of the dish (*see* **Note 18**).

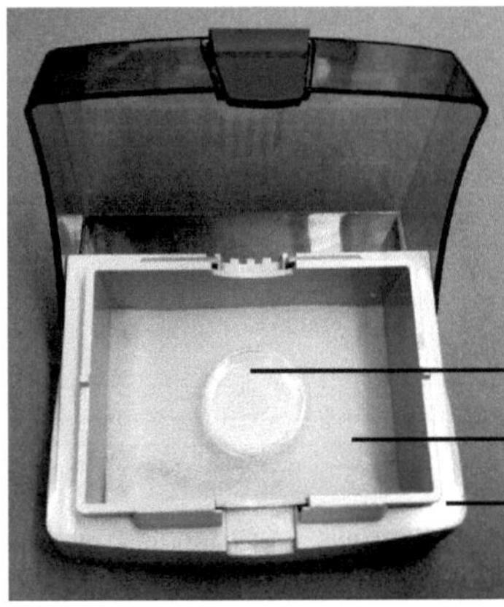

Fig. 2 Humidity chamber for small volume incubations. A moist paper towel was placed on the bottom of a tip box, creating a "humidity chamber" to prevent desiccation of small sample volume used in glass-bottom dishes for staining

9. Incubate the dish in the humidity chamber for 5–10 min at 37 °C.
10. Carefully wash the cells 3× with 300 μL PBS/wash (*see* **Note 19**).
11. Add 300 μL PBS to the dishes and proceed to confocal imaging (*see* **Notes 20**, **21** and **22** and Table 1).
12. Set up three confocal channels to acquire simultaneously in the following order: (1) Crystal reflection, (2) Lysosome fluorescence, (3) Cell surface fluorescence.
13. For crystal reflection, the AOBS® (available in the LEICA confocal instruments) of the first channel has to be set on roughly 10 % laser light to pass the collection channel. In other confocal systems, set the instrument on reflection mode.
14. Capture images as single channels and arrange them as RGB-pictures using Adobe Photoshop (*see* Fig. 3).

4 Notes

1. Bone marrow from C57BL/6 mice was isolated and cultured for 24 h with 10 ng/mL recombinant GM-CSF. Cells were infected with J2 virus, carrying v-myc and v-raf/mil oncogenes, for an additional 24 h. GM-CSF in the medium was reduced by 50 % every 48 h. After 2 weeks, cells will no longer be dependent on growth factor additive in the culture medium.
2. L-leucyl L-leucine methyl ester is a lysosomotropic agent.
3. Bafilomycin A1 is an inhibitor of the vacuolar-type (H+)-ATPase.
4. Optional: Prior to incubation, you may choose to centrifuge the plate for 3 min at 400 × *g*. This will allow the cells to adhere more homogenously.

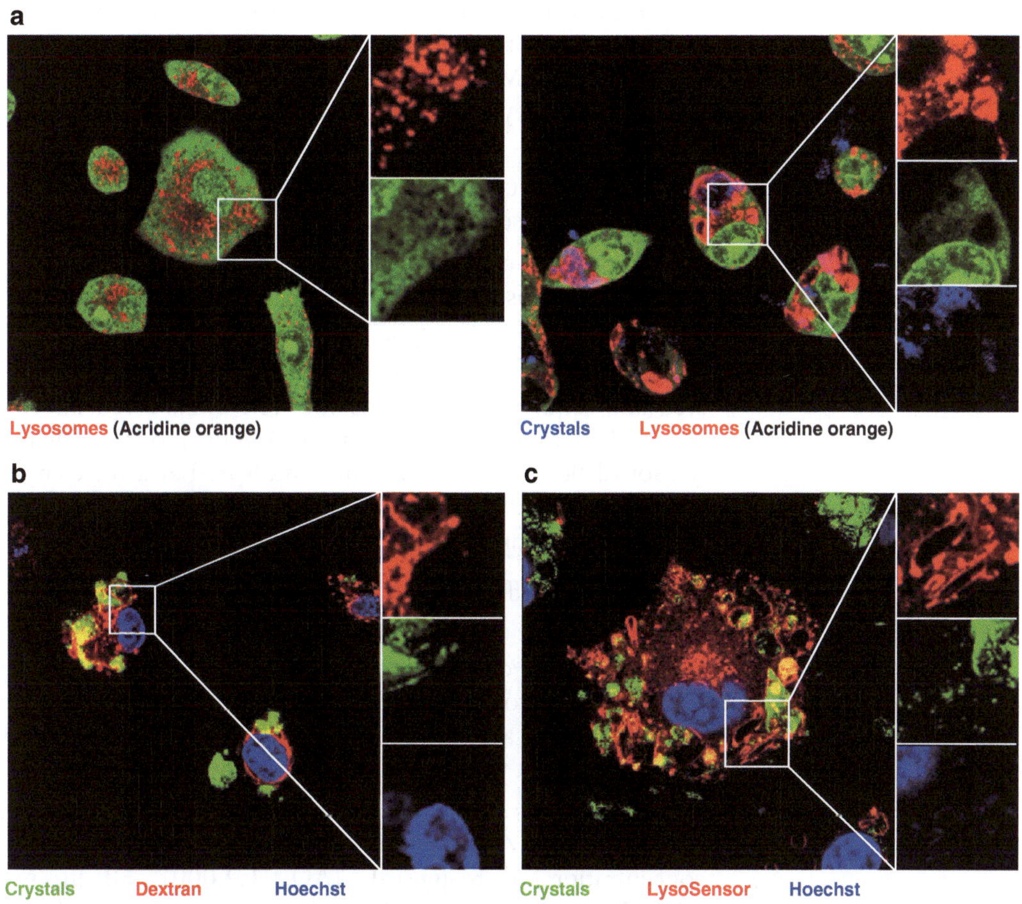

Fig. 3 Confocal laser reflection and fluorescence imaging of immortalized macrophages. Cells labeled with acridine orange (**a**), Alexa-488 dextran (**b**) or with LysoSensor™ Green (**c**) were treated with crystals or left untreated (**a**, *left picture*). Merged pictures of co-localized (*yellow*) crystals (**a**, *blue*: **b** and **c**, *green*) localized in lysosomal compartments (*red*)

5. Optional: At this stage you can choose to use an inhibitor such as bafilomycin (*see* **Note 3**). Dilute the bafilomycin A1 stock in DMEM to obtain a 10× concentration of 1 μM (1:160). Add 50 μL of a 10× bafilomycin A1 solution (1 μM) to each well to obtain a final concentration of 100 nM and incubate the plate for 1 h at 37 °C. Replace the medium with 450 μL of fresh DMEM/well and proceed with the next step of the protocol.
6. 100 nM is a good starting concentration. Since inhibitor concentrations need to be evaluated, e.g., for different cell types, you may need to prepare different dilutions.
7. Dilute the AO stock solution (10 mg/mL) in DMEM to a final concentration of 5 μg/mL (1:2,000).
8. Add by pipetting slowly from the side of the well. Adding reagents too fast will result in dislodging of the cells.

9. Dilute the Silica (50 mg/mL) and Leu-Leu-OMe (1 M) stocks in DMEM to obtain 10× concentrations of Silica (5 mg/mL) and Leu-Leu-OMe (1 μM). For titration you can prepare a serial dilution by e.g., one part 10× reagent to one part DMEM.
10. For kinetics, start with the longest incubation time point (6 h) followed by the second last (i.e., 2 h later for 4 h) and so on.
11. The supernatant can be frozen at −20 °C or directly used, e.g., for cytokine measurements.
12. Acridine orange (AO) binds to nucleic acids and emits green fluorescence. In low pH environments AO becomes protonated and sequestered. In lysosomes AO emits a red fluorescence when excited with a blue laser (488 nm) source. The red lysosomal fluorescence emission is measured at a wavelength between 600 and 650 nm (PE-filter channel).
13. Use a gate to exclude cell debris and aggregates.
14. Expect a loss of fluorescence change of about two log scales.
15. Some cells may require additional treatment of the glass-bottom dishes. For better adherence, glass-bottom dishes can be coated with collagen. For this, prepare 50 μg/mL collagen in sterile ddH_2O and incubate at 37 °C for 1 h. Wash the dishes with PBS and let them dry before use.
16. Dilute the stocks of AO (10 mg/mL), DQ-OVA (5 mg/mL), 488-Dex (5 mg/mL) or LTG (1 mM) in DMEM to a final concentration of 5 μg/mL (AO, 1:2,000), 50 μg/mL (DQ-OVA, 1:100), 20 μg/mL (488-Dex, 1:250) or 50 nM (LysoTracker Green, 1:20,000). All fluorescent dyes indicated here are used to locate lysosomes. AO and LysoTracker are indirect indicators of lysosomal damage (loss of lysosomal fluorescence or color change), whereas DQ-OVA and DEX can be traced following their translocation into the cytosol after lysosomal rupture.
17. Due to the small volume of medium in the glass-bottom dish, care must be taken. The wet paper towel in the tip box will maintain humidity and prevent desiccation of the samples.
18. Dilute the CtxB stock (1 mg/mL) in DMEM to a working concentration of 10 μg/mL (CtxB, 1:100).
19. Stimulated cells tend to be less adherent and are easily flushed away. Carefully add and remove PBS from the side of the dish and avoid adding or removing fluids directly over the area of cell growth.
20. As an optional step, additional staining of nuclei (e.g., with DAPI) can be performed. Be aware, that DAPI's broad emission spectrum (400–550 nm) will require use of sequential imaging.

21. We used the Leica TCS SP5 confocal laser microscope equipped with an Acousto Optical Beam Splitter (AOBS®). The AOBS® allows for combining laser reflection and standard fluorescence imaging.
22. Conjugated lysosomotropic dyes have been selected to emit their spectrum at a sufficient distance to the cell membrane dye (CtxB).
23. Denoted wavelengths are according to the manufacturer's manual. They may vary slightly using different solvents.

References

1. Thiele DL, Lipsky PE (1990) Mechanism of L-leucyl-L-leucine methyl ester-mediated killing of cytotoxic lymphocytes: dependence on a lysosomal thiol protease, dipeptidyl peptidase I, that is enriched in these cells. Proc Natl Acad Sci USA 87(1):83–87
2. Hornung V, Bauernfeind F, Halle A, Samstad EO, Kono H, Rock KL, Fitzgerald KA, Latz E (2008) Silica crystals and aluminum salts activate the NALP3 inflammasome through phagosomal destabilization. Nat Immunol 9(8):847–856. doi:ni.1631 [pii]10.1038/ni.1631
3. Halle A, Hornung V, Petzold GC, Stewart CR, Monks BG, Reinheckel T, Fitzgerald KA, Latz E, Moore KJ, Golenbock DT (2008) The NALP3 inflammasome is involved in the innate immune response to amyloid-beta. Nat Immunol 9(8):857–865. doi:ni.1636 [pii]10.1038/ni.1636
4. Duewell P, Kono H, Rayner KJ, Sirois CM, Vladimer G, Bauernfeind FG, Abela GS, Franchi L, Nunez G, Schnurr M, Espevik T, Lien E, Fitzgerald KA, Rock KL, Moore KJ, Wright SD, Hornung V, Latz E (2010) NLRP3 inflammasomes are required for atherogenesis and activated by cholesterol crystals. Nature 464(7293):1357–1361. doi:nature08938 [pii]10.1038/nature08938

Chapter 4

Assessment of Inflammasome Activation in Primary Human Immune Cells

Theo S. Plantinga, Leo A.B. Joosten, and Mihai G. Netea

Abstract

Inflammasomes are central regulators of inflammation, responsible for cleavage of the inactive pro-inflammatory cytokines IL-1β and IL-18 into their biologically active counterparts. Several regulatory stages within the pathways responsible for the production of these cytokines have been identified. In this chapter, methods are described for assessing the degree of activity of these regulatory stages, which include mRNA transcription, caspase-1 activation, and secretion of the bioactive proteins.

Key words Human primary cells, Inflammasomes, Caspase-1, Quantitative PCR, ELISA, Western blot, Fluorescent activity assay

1 Introduction

Stimulation and release of proinflammatory cytokines is an essential step for the activation of an effective innate host defense, and subsequently for the modulation of adaptive immune responses. Interleukin-1β (IL-1β) and IL-18 are important proinflammatory cytokines that on one hand activate monocytes, macrophages and neutrophils, and on the other hand induce Th1 and Th17 adaptive cellular responses [1, 2]. Interleukin-1β family cytokines are synthesized as inactive precursors, and the processing of pro-IL-1β and pro-IL-18 is dependent on cleavage by proteases [3]. One of the most important proteolytic enzymes processing pro-IL-1β and pro-IL-18 is the cysteine-protease caspase-1, which itself is activated by protein platforms called the inflammasomes [4].

The inflammasomes are multiprotein oligomers consisting of the protease caspase-1, Apoptosis-associated Speck-like protein containing a CARD (ASC) and a pattern-recognition receptor (PRR) of the NOD-like (NLRs) family of receptors. Several inflammasomes have been described for the activation of caspase-1, each of them based on the associated NLRs [5].

Christine M. De Nardo and Eicke Latz (eds.), *The Inflammasome: Methods and Protocols*, Methods in Molecular Biology, vol. 1040, DOI 10.1007/978-1-62703-523-1_4, © Springer Science+Business Media New York 2013

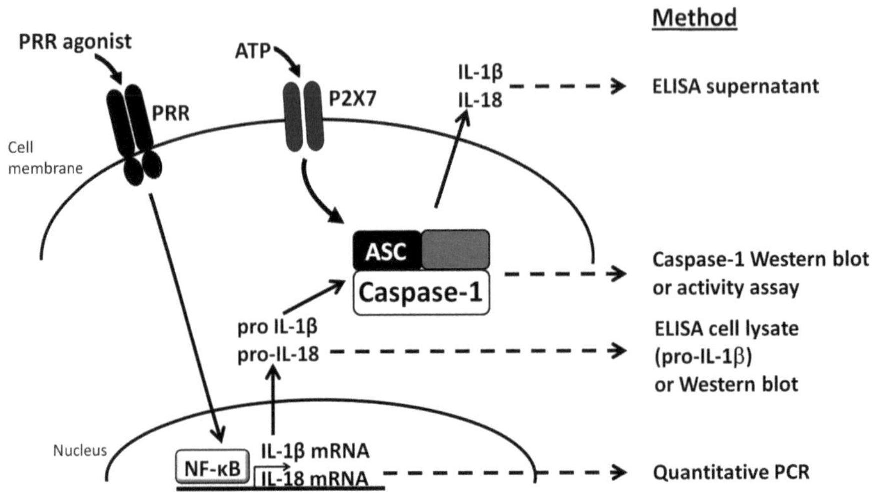

Fig. 1 Schematic overview of the pathways leading to production and secretion of bioactive IL-1β and IL-18 accompanied by the appropriate techniques to assess for every regulatory step in this process. *PRR* pattern-recognition receptor; *ATP* adenosine triphosphate; *ASC* apoptosis-associated Speck-like protein containing a CARD

Through CARD–CARD and pyrin–pyrin domain interactions, a large macromolecular complex is formed to represent a scaffold for the recruitment and activation of pro-caspase-1. It is believed that caspase-1 activation in the inflammasome is induced by formation of oligomers and proximity between caspase-1 molecules.

Inflammasomes are typically expressed in myeloid cells, which include monocytes and macrophages. Recent studies have revealed important differences in the regulation of inflammasome activation between these two cell types, with constitutive caspase-1 activity in monocytes, and inflammasome-inducible activity in macrophages [6]. Subsequently, monocytes require only one stimulus to induce induction of mRNA transcription for pro-IL-1β and pro-IL-18, while processing is automatically performed by the constitutively active caspase-1. In contrast, macrophages require a second stimulus for release of bioactive IL-1β and IL-18 that activates the inflammasome for subsequent caspase-1 activation. One of these signals is ATP signaling through the purinergic P2X7 receptor [7].

Several methods are described below for the assessment of the inflammasome activation pathway in human primary immune cells leading to the secretion of bioactive IL-1β and IL-18. These include: mRNA expression of IL-1β and IL-18, detection of the intracellular pro-IL-1β and pro-IL-18 proteins, measurement of caspase-1 activity, and detection of secreted IL-1β and IL-18 in the supernatant (*see* Fig. 1).

2 Materials

Prepare all solutions using deionized water. Prepare and store all reagents at room temperature, unless otherwise indicated.

2.1 Cell Stimulation Assays (Monocytes and Macrophages)

1. Blood donated by human volunteers as buffycoats or in 10 mL EDTA tubes.
2. Sterile 50 mL tubes.
3. Sterile phosphate buffered saline (PBS) at 4 °C.
4. Centrifuge at 4 °C with adjustable acceleration and deceleration.
5. Ficoll for isolation of the mononuclear cell fraction.
6. Pasteur pipette.
7. 50 mL syringes and long needles (at least 15 cm).
8. Culture medium for up to 48 h of cell culture: RPMI Dutch modification supplemented with 5 mg/mL gentamycin, 100 mM pyruvate, and 200 mM glutamax.
9. Culture medium for up to 7 days of cell culture: RPMI Dutch modification supplemented with 5 mg/mL gentamycin, 100 mM pyruvate, and 200 mM glutamax with the addition of 10 % human pooled serum (*see* **Note 1**).
10. 100 ng/mL GM-CSF diluted in culture medium (*see* **Note 2**).
11. 100 ng/mL M-CSF diluted in culture medium (*see* **Note 2**).
12. Flat or round bottom 96-well plates (*see* **Note 3**).
13. Cell culture incubator maintained at 37 °C and 5 % CO_2.
14. Trizol® (Invitrogen). Store at 4 °C.

2.2 IL-1β Secretion Assay

1. PRR ligand (e.g., lipopolysaccharide (LPS), final concentration 10 ng/mL).
2. Culture medium containing 1 mM adenosine triphospohate (ATP).

2.3 Quantitative PCR for IL-1β and IL-18

1. Water bath at 65 °C.
2. cDNA synthesis kit (Bio-Rad iScript).
3. SYBR Green master mix (Applied Biosystems).
4. IL-1β mRNA expression primers: forward 5′-CAG-CTA-CGA-ATC-TCC-GAC-CAC-3′; reverse 5′-GGC-AGG-GAA-CCA-GCA-TCT-TC-3′ (*see* **Note 4**).
5. IL-18 mRNA expression primers: forward 5′-TGT-CGC-AGG-AAT-AAA-GAT-GGC-T-3′; reverse 5′-CCT-TGG-TCA-ATG-AAG-AGA-ACT-TGG-T-3′ (*see* **Note 4**).
6. Housekeeping gene β2-microglobulin mRNA expression primers: forward 5′-ATG-AGT-ATG-CCT-GCC-GTG-TG-3′; reverse 5′-CCA-AAT-GCG-GCA-TCT-TCA-AAC-3′ (*see* **Note 4**).

2.4 ELISA for Uncleaved Pro-IL-1β and Cleaved Mature IL-1β and IL-18

1. ELISA kits for total IL-18, total IL-1β, or pro-IL-1β (R&D Systems).
2. Automated plate washer (or wash bottles).

2.5 Western Blot for Detection of the Pro-form and Cleaved Form of (Pro-) Caspase-1, (Pro-)IL-1β and (Pro-)IL-18

1. Lysis buffer: 50 mM Tris, pH 7.4, 150 mM NaCl, 2 mM ethylenediaminetetraacetic acid, 2 mM ethyleneglycoltetraacetic acid, 10 % glycerol, 1 % Triton X-100, 40 mM α-glycerophosphate, 50 mM sodium fluoride, 200 μM sodium vanadate, 10 μg/mL leupeptin, 10 μg/mL aprotinin, 1 μM pepstatin A, and 1 mM phenylmethylsulfonyl fluoride.
2. Heating block at 95 °C.
3. 5× Laemmli buffer containing 1 M dithiothreitol (DTT): 200 mM Tris–HCl, pH 6.8, 50 % glycerol, 0.0625 % Bromophenol blue, 10 % SDS, in H_2O, plus 1 M DTT (*see* **Note 5**).
4. Sodium dodecyl sulfate–polyacrylamide gel electrophoresis (SDS-PAGE), 10 and 15 % polyacrylamide gels.

   For a 10 % running gel: Mix 3 mL 40 % acrylamide/Bis with 6 mL deionized water, 3 mL 1.5 M Tris–HCl, pH 8.8, 130 μL 10 % SDS in H_2O, 100 μL 10 % ammonium persulfate in H_2O, and 30 μL TEMED.

   For a 15 % running gel: Mix 4.5 mL 40 % acrylamide/Bis with 4.5 mL deionized water, 3 mL 1.5 M Tris–HCl, pH 8.8, 130 μL 10 % SDS in H_2O, 100 μL 10 % ammonium persulfate in H_2O, and 30 μL TEMED.

   For the stacking gel: Mix 400 μL 40 % acrylamide/Bis with 2.5 mL deionized water, 1.0 mL 0.5 M Tris–HCl, pH 6.8, 40 μL 10 % SDS in H_2O, 25 μL 10 % ammonium persulfate in H_2O, and 20 μL TEMED.
5. Nitrocellulose membrane (0.2 μm).
6. Wash buffer: 5 % bovine serum albumin/Tris-buffered saline/Tween 20 (5 % BSA/TBS-T).
7. Block solution: 5 % (wt/vol) milk powder dissolved in TBS-T.
8. Antibody incubation solution: 5 % BSA/TBS-T.
9. Primary antibodies:
   (a) Anti-human caspase-1 p10 (sc-515, Santa Cruz Biotechnology).
   (b) Anti-human caspase-1 p45 (sc-622, Santa Cruz Biotechnology).
   (c) Anti-human IL-1β (Cell Signaling Technology).
   (d) Anti-human pro-IL-18 and anti-human IL-18 (R&D Systems).
   (e) Anti-human β-actin (Sigma).

10. Horseradish peroxidase-conjugated goat anti-rabbit or rabbit anti-mouse antibody (AbD Serotec).
11. Densitometer.
12. Molecular Analyst software (Bio-Rad).

2.6 Measurement of Caspase-1 Enzyme Activity

1. Caspase-1 Fluorometric Assay Kit (Biovision).
2. FLUOstar Optima Microplate (BMG Labtech).

3 Methods

3.1 Cell Stimulation Assays (Monocytes and Macrophages)

1. Isolate primary human immune cells from buffycoats or EDTA tubes. To do this, add a maximum of 6 mL (in case of buffycoats) or 20 mL (in case of EDTA tubes) of blood into a 50 mL tube, and make up to 35 mL with sterile PBS.
2. Slowly add 15 mL of Ficoll to the bottom of the 50 mL tube, by using a syringe and a long needle that reaches the bottom of the tube (*see* **Note 6**).
3. Obtain the mononuclear cell fraction by density centrifugation with Ficoll (20 min, 1,600 ×*g*, no brake and slow acceleration at room temperature).
4. Transfer the white cells representing the middle layer on top of the Ficoll into another tube with a Pasteur pipette (*see* **Note 7**).
5. Wash 2× with sterile cold PBS (*see* **Note 8**), centrifuging for 10 min at 1,600 ×*g* at 4 °C and discarding supernatant between each wash (*see* **Note 9**).
6. Reconstitute the cells in culture medium after the final wash and adjust the volume to reach the desired concentration (*see* **Note 10**).
7. Add the cells to flat or round bottom (*see* **Note 3**) 96-well plates in a volume of 100 μL and add 100 μL of the stimulus in the appropriate concentration dissolved in culture medium.
8. Incubate the plate at 37 °C for 1 h and then discard the non-adherent lymphocytes (*see* **Note 11**).
9. Add 100 μL fresh culture medium and incubate the cells for example for 24 h at 37 °C, with or without a stimulus such as lipopolysaccharide (LPS) with a final culture volume of 200 μL.
10. Centrifuge the plate at 1,500 ×*g* for 5 min (*see* **Note 12**) and collect the supernatants for assessment of cytokine secretion by specific ELISAs (*see* Subheading 3.4).
11. Add 200 μL of Trizol® to each well, to lyse the cells for RNA extraction (*see* **Note 13**).

3.2 IL-1β Secretion Assay

To determine the capacity of cells to secrete IL-1β, stimulate cells with a PRR ligand (e.g., LPS) together with adenosine triphosphate (ATP), which binds to the purinergic P2X7 receptor (*see* Fig. 1).

1. Stimulate monocytes from **step 8** of Subheading 3.1, with the activating stimulus for 3 h at 37 °C. In the case of LPS, incubate the cells with 10 ng/mL LPS for 3 h at 37 °C in round (in case of complete mononuclear cell pool) or flat (in case of enrichment for monocytes) bottom 96-well plates in a total volume of 200 μL/well (*see* **Note 3**).
2. Centrifuge the plate at 1,500 × *g* for 5 min and collect the supernatants for IL-1β ELISA measurements (*see* **Note 12**) by ELISA as described in Subheading 3.4.
3. Resuspend the cell pellet in 200 μL/well fresh culture medium containing 1 mM ATP and incubate for 30 min at 37 °C.
4. Centrifuge the plate at 1,500 × *g* for 5 min and collect the supernatants and assess for IL-1β secretion by ELISA as described in Subheading 3.4.

3.3 Quantitative PCR for IL-1β and IL-18

1. Isolate RNA from the samples obtained at **step 11** in Subheading 3.1 according to the Invitrogen Trizol® RNA isolation protocol.
2. Store RNA at −80 °C until use for cDNA synthesis.
3. Synthesize cDNA by using a cDNA synthesis kit (*see* **Note 14**).
4. Perform Real Time-PCR with SYBR Green to determine IL-1β and IL-18 mRNA expression using the primers stated in **steps 4** and **5** of Subheading 2.3. Use a housekeeping gene as a background reference (e.g., β2-microglobulin).

3.4 ELISA for Uncleaved Pro-IL-1β and Cleaved Mature IL-1β and IL-18

For the measurement of cytokine concentrations in the supernatant (secretion) or in the cell lysate (produced but uncleaved pro-IL-1β), use commercially available ELISA (enzyme-linked immunosorbent assay) kits that are specific for total IL-18, total IL-1β, or pro-IL-1β (*see* **Note 15**). No kits for pro-IL-18 are currently available. For measurement of intracellular pro-IL-1β:

1. Add new culture medium to the cells after removing the cell supernatants as described in **step 10** of Subheading 3.1 and **step 4** of Subheading 3.2.
2. Subject the cells to 3–5 freeze thaw cycles to lyse the cells and release the produced pro-IL-1β from the cell cytoplasm.
3. Carry out ELISA according to manufacturer's protocols.

3.5 Western Blot for Detection of the Pro-form and Cleaved Form of (Pro-) Caspase-1, (Pro-) IL-1β, and (Pro-)IL-18

In order to detect and quantify the amount of pro-caspase-1, cleaved caspase-1, pro-IL-18, cleaved IL-18, pro-IL-1β, and cleaved IL-1β; these proteins can be discriminated by protein size and can be stained by Western blotting.

1. For Western blotting, lyse 10×10^6 cells in 100 μL lysis buffer (*see* **Note 16**).
2. Freeze and thaw the homogenate once and centrifuge at 4 °C for 10 min at 14,000 × *g*, and collect the supernatant for analysis by Western blot.
3. Add 25 μL 5× Laemmli buffer containing DTT to the lysates and boil the samples in a 95 °C heating block for 10 min (*see* **Note 5**).
4. Measure protein concentration in the cell lysates using a BCA protein assay kit. Load up to 50 μg of protein per well on the SDS-PAGE gel.
5. Subject equal amounts of protein to sodium dodecyl sulfate–polyacrylamide gel electrophoresis (SDS-PAGE) using 10 and 15 % polyacrylamide gels at a constant voltage of 100 V for ~ 2 h. Stop the running of the gel when the blue color of the Laemmli buffer has reached the bottom of the gel.
6. After SDS-PAGE, transfer the proteins to nitrocellulose membrane (0.2 μm).
7. Block the membrane with 5 % (wt/vol) milk powder dissolved in TBS-T for 1 h at room temperature with agitation.
8. Incubate the membrane overnight with the primary antibody at 4 °C with agitation; anti-human caspase-1 p10 (dilution 1:1,000), anti-human caspase-1 p45 (dilution 1:1,000), anti-human IL-1β (dilution 1:1,000), anti-human pro-IL-18 (dilution 1:1,000), anti-human IL-18 (dilution 1:1,000), or anti-human β-actin (dilution 1:1,000) in 5 % BSA/TBS-T (*see* **Note 17**).
9. Wash the blots 3× in TBS-T with agitation at room temperature (15 min each wash).
10. Incubate with HRP-conjugated goat anti-rabbit or rabbit anti-mouse antibody at a dilution of 1:10,000 in 5 % (wt/vol) milk powder dissolved in TBS-T for 1 h with agitation at room temperature (*see* **Note 17**).
11. Wash the blots 3× with TBS-T as in **step 9**.
12. Develop the blots with chemiluminescence according to the manufacturer's instructions.
13. Quantify protein expression by densitometry and signal analysis using Molecular Analyst software.
14. Calculate the ratio between the intensity of the protein of interest and β-actin to assess for differences in expression of the

protein of interest after normalization based on the β-actin staining (*see* **Note 18**).

3.6 Measurement of Caspase-1 Enzyme Activity (See Note 19)

1. Stimulate 2 ×10^6 monocytes or monocyte derived (M1 or M2) macrophages for 6 h at 37 °C with an activating stimulus such as LPS, *see* **step 9** of Subheading 3.1 (*see* **Note 11**).
2. Lyse cells with 50 μL cell lysis buffer on ice for 10 min.
3. Add equal volumes (50 μL) of 2× reaction buffer and 5 μL of the YVAD-AFC substrate to the cell lysate in a 96-well plate and incubate for 1–2 h at 37 °C.
4. Detect the emitted fluorescence by FLUOstar Optima Microplate at 90 s intervals for 200 cycles using 400 nm excitation and 505 nm emission filters.
5. Calculate the relative activity (linear fold increase) as compared to the untreated negative control.

4 Notes

1. Obtain human pooled serum from healthy volunteers that have donated blood collected in serum tubes. Centrifuge these tubes and mix the supernatant with serum supernatants from other healthy volunteers. Aliquot the serum into 2 mL tubes and store at −80 °C until use.
2. Reconstitute recombinant proteins M-CSF and GM-CSF according to the manufacturers' instructions. After reconstitution, prepare small volume aliquots and store at −80 °C until use.
3. Round bottom plates are used when monocytes do not need to be separated from the other mononuclear cells in the fraction. If this selection step is required, use flat bottom plates as this will enable the monocytes to attach to the plastic (the other mononuclear cells such as T and B cells do not attach) and, if desired, to differentiate into macrophages.
4. Stock concentrations of primers should be kept at 100 μM, and stored at −20 °C. Final concentration in the PCR mix is 10 μM.
5. Add dithiothreitol (DTT) to the Laemmli buffer in order to reduce disulphide bridges in proteins, as this will adopt the random-coil configuration necessary for separation by size.
6. When adding the Ficoll below the PBS diluted blood in 50 mL tubes with a syringe and a long needle, make sure the Ficoll is ejected slowly from the syringe in order to prevent mixing of the PBS diluted blood with the Ficoll. The Ficoll should become a separate layer of clear liquid beneath the PBS diluted blood before starting the centrifugation step.

7. Some cells that are not on top of the Ficoll but stick to the wall just above the Ficoll layer can be tapped off the wall for use in the experiment as well.
8. Ensure cold PBS is used in wash steps, as this will prevent the cells sticking to the inside wall of the tubes.
9. Before adding new PBS to the cells during the washing steps, loosen the cells from the bottom of the tubes by tapping the tube. This will enable the cells to be washed more efficiently and to completely remove any remaining Ficoll.
10. The cells can be counted by using a counting chamber under the microscope or by an automated cell counter. In 96-well plates the number of cells to be added can vary to up to one million cells in a total volume of 100 μL. Therefore, adjust the cell concentration to a maximum of ten million cells per mL.
11. Alternatively, cytokine stimulation can be performed in differentiated macrophages. For differentiation of monocytes towards macrophages, incubate monocytes, after removal of the non-adherent lymphocytes, for 7 days with culture medium supplemented with 10 % human serum. Alternatively, for directing the monocytes into a macrophage type I (M1) or macrophage type II (M2) differentiation, add 100 ng/mL GM-CSF or 100 ng/mL M-CSF to the culture medium, respectively, and incubate for 7 days. M1 and M2 macrophages are distinct cell types, with M1 as the pro-inflammatory phenotype and the M2 as the anti-inflammatory tissue-remodeling phenotype. As M1 and M2 macrophages are of importance in the initiation and termination of the immune response, respectively, both macrophage types are of great interest with respect to their response to particular (e.g., microbial) stimuli.
12. This centrifugation step pellets cellular debris and intact cells, excluding them from the supernatants.
13. Add up to 200 μL of Trizol® to every well and store the plate at −80 °C until RNA isolation.
14. After RNA isolation and reconstitution of the RNA in rehydration solution, measure the RNA concentration (e.g., by NanoDrop). Use up to 1 μg of RNA for cDNA synthesis within a total volume of 20 μL [consisting of 15 μL RNA (+H_2O), 1 μL reverse transcriptase, and 4 μL reaction mix].
15. Dilute the supernatants before performing the ELISA, preferably 10 times and 100 times, to maximize the detection window of the standard curve.
16. For easy collection of the cell lysate for Western blot, culture the cells in 1.5 mL tubes for this experiment. Following incubation, centrifuge the tubes at 14,000 × g, remove the supernatant, and lyse the cells.

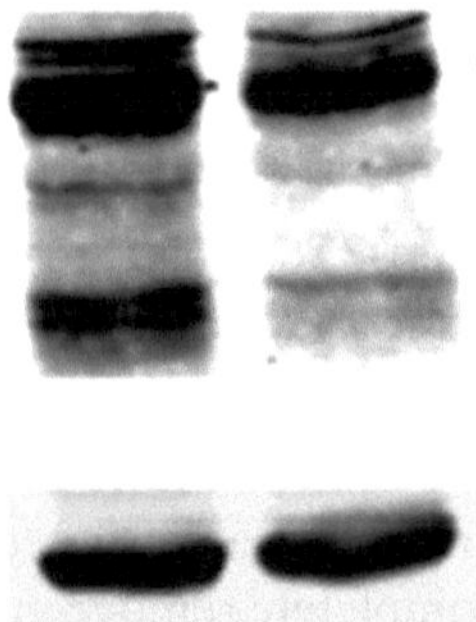

Fig. 2 Assessment of caspase-1 protein expression, both the inactive p45 pro-form and the active p35 protein. Actin served as a loading control

17. To minimize the volume needed for the incubation with the primary and secondary antibody, blots can be incubated in 50 mL conical tubes, aligned along the wall of the tube. In this way, the incubation volume can be reduced to 5 mL.
18. The degree of caspase-1 activation can be assessed by calculating the ratio between the p35 (active caspase-1) and p45 (inactive pro-caspase-1) fragments (*see* Fig. 2).
19. An alternative method: The Caspase-1/ICE Fluorometric Assay Kit provides a simple and convenient means for assaying the activity of caspase-1 by its recognition of the sequence YVAD. The assay is based on detection of cleavage of substrate YVAD-AFC (AFC: 7-amino-4-trifluoromethyl coumarin). YVAD-AFC emits blue light (400 nm); upon cleavage of the substrate by caspase-1, free AFC emits a yellow-green fluorescence (505 nm), which can be quantified using a fluorometer or a fluorecence microtiter plate reader. Comparison of the fluorescence of AFC from a treated sample with an untreated control allows determination of the fold increase in caspase-1 activity.

Acknowledgments

This work was supported by a Veni and a Vici grant of the Dutch Organization for Scientific Research (NWO) to Theo S. Plantinga and Mihai G. Netea, respectively.

References

1. Arend WP, Palmer G, Gabay C (2008) IL-1, IL-18, and IL-33 families of cytokines. Immunol Rev 223:20–38
2. Dinarello CA (2006) Interleukin 1 and interleukin 18 as mediators of inflammation and the aging process. Am J Clin Nutr 83(2):447S–455S
3. van de Veerdonk FL, Netea MG, Dinarello CA, Joosten LA (2011) Inflammasome activation and IL-1beta and IL-18 processing during infection. Trends Immunol 32(3):110–116. doi:S1471-4906(11)00004-4 [pii] 10.1016/j.it.2011.01.003

4. Gross O, Thomas CJ, Guarda G, Tschopp J (2011) The inflammasome: an integrated view. Immunol Rev 243(1):136–151
5. Martinon F, Tschopp J (2004) Inflammatory caspases: linking an intracellular innate immune system to autoinflammatory diseases. Cell 117(5):561–574
6. Netea MG, Nold-Petry CA, Nold MF, Joosten LA, Opitz B, van der Meer JH, van de Veerdonk FL, Ferwerda G, Heinhuis B, Devesa I, Funk CJ, Mason RJ, Kullberg BJ, Rubartelli A, van der Meer JW, Dinarello CA (2009) Differential requirement for the activation of the inflammasome for processing and release of IL-1beta in monocytes and macrophages. Blood 113(10):2324–2335
7. Ferrari D, Pizzirani C, Adinolfi E, Lemoli RM, Curti A, Idzko M, Panther E, Di VF (2006) The P2X7 receptor: a key player in IL-1 processing and release. J Immunol 176(7):3877–3883

Chapter 5

NLRP3 Inflammasome Activation and Cytotoxicity Induced by Particulate Adjuvants

Marie Yang, Claire H.A. Hearnden, Ewa Oleszycka, and Ed C. Lavelle

Abstract

The ability of particulate materials to provoke inflammatory immune responses has been well documented. In the case of endogenous and environmental particulates, these effects can often lead to pathological disorders. In contrast, particulate adjuvants incorporated into vaccines promote immune responses, which in turn provide efficient protection against infectious diseases. In recent years, studies have revealed that the NLRP3 inflammasome plays a key role in particulate-driven inflammation and its associated cytotoxicity. Hence, this chapter covers protocols useful to (1) assess NLRP3 inflammasome activation triggered by particulate adjuvants or materials in mouse bone marrow-derived dendritic cell (BMDCs) differentiated cultures, and (2) measure particle-induced cytotoxicity. More specifically, protocols are described for the preparation and differentiation of BMDCs, their priming and stimulation using particulate NLRP3 agonists such as monosodium urate monohydrate (MSU) and the vaccine adjuvant alum. We then detail protocols to assess particulate-driven cytotoxicity via flow cytometry using annexin V-propidium iodide (PI) and novel dye LIVE/DEAD® aqua stain. General considerations are provided that warn against the use of endotoxin-contaminated particles and emphasize the use of experimental controls. Suggestions are also outlined for further assessment of the immunomodulatory effects of particulate materials in vivo using the mouse peritonitis model.

Key words Particulate adjuvants, NLRP3 inflammasome, Cytotoxicity, IL-1β, Caspase-1

1 Introduction

Exposure to particulate materials can provoke inflammatory responses and exert cytotoxic effects. Chronic environmental exposure to particles such as silica crystals and asbestos fibers, for example, can cause deposits in the lung leading to mesothelioma [1, 2]. Similarly, endogenous particulate stimuli, such as uric acid crystals, resulting from oxidative stress, injury or abnormal metabolism, may lead to the development of inflammatory and autoimmune disorders. On the other hand, particulate vaccine adjuvants, particularly alum salts, play a vital role in the promotion of protective immune responses in subunit vaccines. As part of the innate

Christine M. De Nardo and Eicke Latz (eds.), *The Inflammasome: Methods and Protocols*, Methods in Molecular Biology, vol. 1040, DOI 10.1007/978-1-62703-523-1_5, © Springer Science+Business Media New York 2013

immune system, inflammasomes govern the processing and secretion of caspase-1 and the pro-inflammatory cytokines IL-1β and IL-18. NLRP3 is currently one of the best-characterized inflammasomes. It consists of three main protein components, the Nod-like receptor (NLR)-family protein NLRP3, pro-caspase-1, and the apoptosis-associated speck-like protein containing a CARD (ASC) adapter, which bridges interactions between the former two. Upon inflammasome assembly, auto-activation of caspase-1 results in cleavage of IL-1β, and subsequent secretion of its biologically active form. The NLRP3 inflammasome is unique in its induction, as it is activated by a range of ligands that are highly structurally diverse [3, 4]. Here, we place the emphasis on particulate materials. The first evidence that particulate materials could activate the NLRP3 inflammasome came from investigations on the inflammatory properties of monosodium urate monohydrate (MSU) crystals [5], the etiologic agent of gout. Using primary cultures, silica crystals and asbestos fibers were also shown to activate the NLRP3 inflammasome, and in vivo, mice deficient for NLRP3 exhibited reduced pulmonary inflammation [4, 6]. Subsequently, particles with adjuvant activity such as alum and chitosan [7–9], and likewise poly(lactide-co-glycolide) (PLG) and polystyrene microparticles [10, 11], were also reported as strong activators of the NLRP3 inflammasome. Recent studies subsequently revealed that NLRP3 played a critical role in adjuvant driven cellular immunity [8] and as such, exploitation of this pathway by vaccines may enhance efficacy, thus reinforcing the importance of investigating NLRP3 activation and understanding the underlying mechanisms. Hence, this chapter details methods to assess particle-induced NLRP3 inflammasome activation in mouse bone marrow-derived dendritic cells (BMDCs).

Assessing the effects of particulate compounds on the viability of cells grown in culture is a common predictor of potential toxic effects in target tissues or whole organisms. In this chapter, we thus also describe protocols to measure the cytotoxic responses classically associated with particulate stimuli. There are a number of assays to choose from, which are primarily based on either measurement of metabolic activity or DNA fragmentation or plasma membrane integrity. In this chapter, we describe two easy-to-perform binding assays: (a) an improved methodology based on the classical AnnexinV-PI staining protocol and (b) the use of a novel generation of amine-reactive fluorescent dyes exemplified by LIVE/DEAD® (Invitrogen) aqua stain. Both represent sensitive, reliable, and convenient methods to distinguish viable cells from necrotic or apoptotic cells based on the principle on plasma membrane integrity (*see* **Notes 1–3**).

2 Materials

All solutions should be prepared using deionized water and brought to the appropriate pH using hydrochloric acid or sodium hydroxide. Store all solutions at room temperature (RT) unless otherwise specified.

2.1 Measurement of Inflammasome Activation

2.1.1 Harvesting and Differentiation of Mouse Bone Marrow Cells into BMDCs

1. Mice (*see* **Note 4**).
2. Ice.
3. A cell culture facility equipped with a 37 °C, 5 % CO_2 water-jacketed cell culture incubator, a laminar flow hood, and a centrifuge capable of holding cell culture plates and tubes.
4. Flushing medium: RPMI 1640, supplemented with 8–10 % Fetal Calf Serum (FCS), 2 mM l-glutamine, 100 IU/mL penicillin, and 100 μg/mL streptomycin.
5. Scissors and forceps.
6. Paper towels.
7. 5, 10, and 25 mL serological pipettes.
8. Pipette filler.
9. 70 % Ethanol.
10. Sterile 50 mL centrifuge tubes.
11. 100 μm cell strainers for use with 50 mL centrifuge tubes.
12. 2 and 20 mL syringes.
13. 19G and 27G needles.
14. Solution of 0.88 % Ammonium Chloride (*see* **Note 5**), filter-sterilized using a 0.22 μm syringe filter.
15. 0.4 % Trypan blue solution.
16. Hemocytometer.
17. BMDC differentiation medium (BMDC Medium): RPMI 1640 supplemented with 8–10 % FCS, 2 mM l-glutamine, 100 IU/mL penicillin, and 100 μg/mL streptomycin, and 20 ng/mL of recombinant murine GM-CSF (*see* **Note 6**).
18. 0.02 % EDTA solution.
19. Sterile, non-tissue culture grade 10 cm petri dishes.

2.1.2 Priming and Stimulation of BMDCs

1. 8- or 12-Tip multichannel pipette with a volume of 50–200 μL per tip.
2. Micropipettes.
3. Sterile pipette tips: 2–200 μL volume and 200–1,000 μL volume.
4. 96-well flat-bottom cell culture plates.

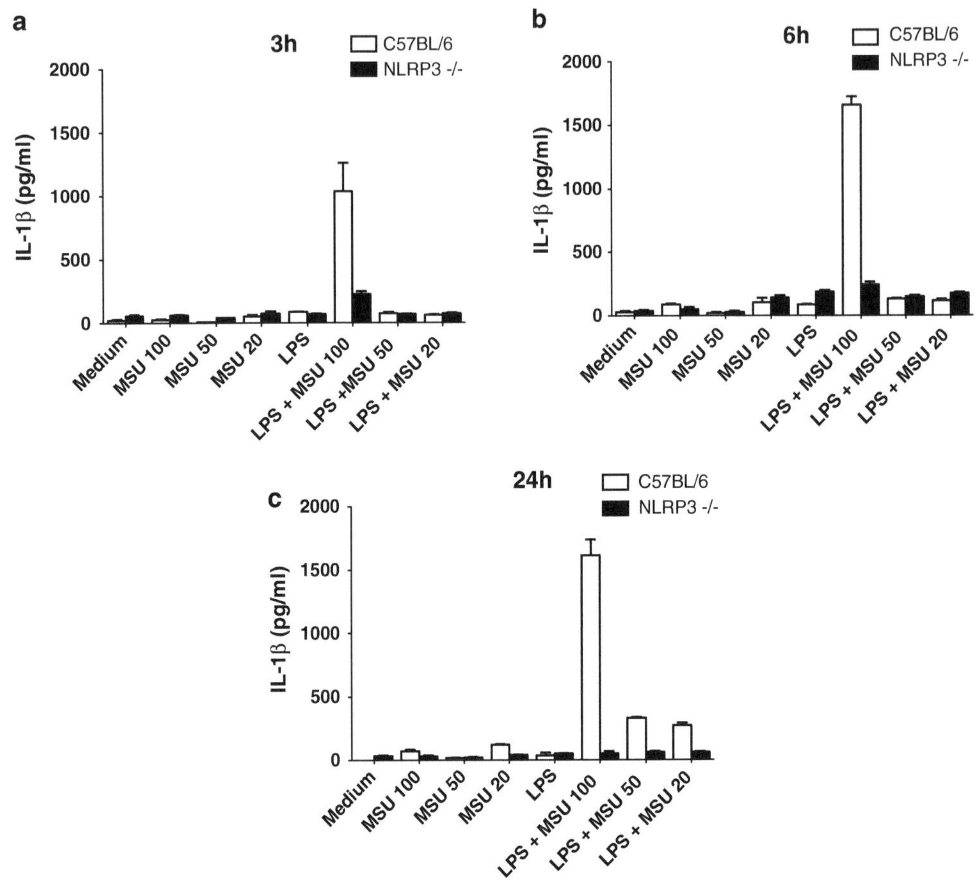

Fig. 1 Kinetics of IL-1β secretion. BMDCs from C57BL/6 and NLRP3−/− mice were left untreated or primed with LPS for 2 h prior to exposure with MSU at 20, 50, and 100 μg/mL for (**a**) 3, (**b**) 6, and (**c**) 24 h. Cell culture supernatants were collected, and IL-1β ELISA (R&D Systems) was performed. All values are represented as mean +/− SD

5. 100 μg/mL *Escherichia coli* K14 ultrapure Lipopolysaccharide (LPS).
6. Inflammasome activators: for example 1 M Adenosine-5′-triphosphate (ATP), 5 mM aluminum hydroxide (e.g., clinically approved Alhydrogel® formulation), or 40 mg/mL of silica.
7. Priming medium: same composition as BMDC differentiation medium (*see* Subheading 2.1.1).
8. 1× sterile PBS.

2.1.3 ELISA Analysis

This section emphasizes the use of ELISA kits for the detection of IL-1β (*see* Fig. 1) and IL-18 as these are processed by caspase-1. However, other cytokines such as IL-6, TNF-α, or IL-12 may also be quantified in the culture supernatants (*see* **Note 7**).

1. High binding flat-bottomed ELISA plates.
2. 1× PBS: 80.0 g NaCl, 14.4 g Na_2HPO_4, 2.4 g KH_2PO_4, 2.0 g KCl. Add deionized H_2O up to 10 L and pH to 7.2 with HCl.

3. ELISA wash buffer: 1× PBS with 0.05 % Tween-20.
4. ELISA dilution buffer: 1× PBS with 1 % BSA.
5. ELISA blocking buffer: 1× PBS with 3 % BSA.
6. Stop solution: 2N sulfuric acid.
7. 5, 10, and 25 mL serological pipettes.
8. Pipette filler.
9. Paper towel.
10. Washbasin or 96-well automatic microplate washer.
11. 8- or 12-tip multichannel pipette with a volume of 50–200 μL per tip.
12. 2–200 μL volume pipette tips.
13. Standard, capture antibody, detection biotin-labelled antibody and avidin-HRP conjugate for cytokine/chemokine of choice.
14. HRP substrate *o*-Phenylenediamine (dihydrochloride) (OPD) solution: 50 mL phosphate citrate buffer (10.19 g anhydrous citric acid, 14.6 g Na_2HPO_4, make up to 1 L with deionized water, pH 5), 20 mg *o*-Phenylenediamine tablet (Sigma), 14 μL Hydrogen peroxide added just before use.
15. 96-well microplate reader.

2.1.4 Western Blot Analysis

1. BMDC differentiated cultures, as described in Subheading 3.1.1.
2. Stimulation medium: RPMI 1640 medium containing less than 5 % FCS.
3. 6- and 12-well tissue culture plates.
4. Micropipettes.
5. Pipette tips: 2–200 μL volume and 200–1,000 μL volume.
6. Toll-like Receptor (TLR) ligands, e.g., LPS at 100 ng/mL (to be used at a final concentration of 1 ng/mL with BMDC cultures).
7. 1× PBS.
8. Particulates of choice: for example aluminum hydroxide (e.g., clinically approved Alhydrogel® formulation) or biodegradable microparticles (e.g., silicon, PLGA).
9. 15 mL centrifuge tubes.
10. Strataclean® beads (Agilent).
11. Vortex.
12. 1.5 mL tubes.
13. Heating block.
14. 15 % polyacrylamide gel (*see* **Note 8**).
15. 1× Laemmli sample buffer: 62.5 mM Tris–HCl, pH 6.8, 25 % glycerol, 2 % SDS, 0.01 % Bromophenol Blue, 5 % Beta-mercaptoethanol (or alternatively use the Biorad Laemmli Sample buffer (Biorad) with the addition of 5 % beta-mercaptoethanol).

16. Western blot transfer buffer: 25 mM Tris, 192 mM glycine, 20 % methanol v/v.
17. Polyvinylidene Difluoride (PVDF) membrane (*see* **Note 9**).
18. Prestained protein molecular weight markers (*see* **Note 10**).
19. Tris-buffered Saline with Tween-20 (TBST): TBS (25 mM Tris, 150 mM NaCl, 2 mM KCl, pH 7.4) and 0.1 % Tween-20.
20. Blocking solution: 5 % nonfat milk in TBST.
21. Rabbit anti-mouse caspase-1 primary antibody (Santa Cruz Labs).
22. HRP-conjugated anti-Rabbit IgG (Jackson ImmunoResearch Lab).
23. Western Blot Stripping buffer (*see* **Note 11**).
24. Goat anti-mouse IL-1β (R&D Systems).
25. HRP-conjugated anti-Goat (Jackson ImmunoResearch Lab).
26. Enhanced chemiluminescence (ECL) substrate (*see* **Note 12**): Millipore ECL detection reagent A + B solutions or home-made reagents A and B mixed at a ratio of 1:1. For reagent A: 0.5 mL luminal, 0.22 mL *p*-coumaric acid, 5 mL 1 M Tris–HCl, pH 8.5, make up to 50 mL with distilled water. For reagent B: 5 mL 1 M Tris–HCl, pH 8.5, 30.5 μL 30 % H_2O_2, make up to 50 mL with deionized water.
27. X-ray films.
28. Developer machine with developing and fixation solutions.
29. Dark room.

2.1.5 Determination of ASC Oligomerization Using Confocal Bioimaging of Live THP-1 Cells

1. THP-1-ASC-GFP cell line [12].
2. DMEM High Glucose Glutamax™ (Invitrogen) medium supplemented with 10 % FCS, 100 IU/mL penicillin and 100 μg/mL streptomycin.
3. DMEM medium containing phorbol 12-myristate-13-acetate (PMA) at 0.5 mM.
4. 35 mm cover glass bottom culture dishes.
5. Hoechst 33342 (for nuclear staining) and CellMask™ deep red (for plasma membrane staining).
6. Confocal microscope equipped with temperature and CO_2-controlled chamber.
7. Particulate stimuli of choice, e.g., aluminum hydroxide (e.g., clinically approved Alhydrogel® formulation), biodegradable particles (e.g., silicon or PLGA).
8. Imaging software for further image processing, e.g., Image *J* software (National Institute of Health, USA).

2.2 Measuring Cytotoxicity

2.2.1 Annexin-Binding Assay in BMDC Cultures

1. BMDC differentiation medium, as described in Subheading 2.1.1.
2. BMDC differentiated cultures (*see* **Note 13** and Subheading 3.1.1)
3. T-175 tissue culture flasks.
4. 96-well or 24-well tissue culture plates (*see* **Note 14**).
5. Hemocytometer or counting chamber.
6. Annexin V-FITC solution: e.g., Promega or homemade (*see* **Notes 15** and **16**) at a stock concentration of 2.5 mg/mL.
7. PI stock solution: we recommend preparing a 1 mg/mL stock solution in PBS and storing at 4 °C protected from light for up to 1 year. Use at a final concentration of 10 μg/mL (e.g., 1 μL in 100 μL of sample), just prior to analysis.
8. 10× concentrated annexin V-binding buffer: 10 mM of HEPES-NaOH, pH 7.4, 150 mM NaCl, 1.5 mM KCl, 1 mM $MgCl_2$, 1.8 mM $CaCl_2$. Store at 4 °C for up to 1 year. Prepare 1× annexin V-binding buffer by making up a 1:10 dilution with distilled water just prior to use (*see* **Note 17**).
9. Necrosis/apoptosis-inducing reagents, e.g., actinomycin-D, cycloheximide, staurosporine, or others, diluted at a concentration deemed optimal.
10. LIVE/DEAD® (Invitrogen) fixable aqua dead cell dye.
11. Cell fixative (*see* **Note 18**).
12. 1× PBS solution.
13. FACS buffer: PBS buffer supplemented with 2 % FCS and 0.05 % sodium azide.
14. FACS analysis tubes compatible with flow cytometer.
15. Flow cytometer and analytical software, e.g., FlowJo® (TreeStar).

3 Methods

3.1 Quantification of Inflammasome Activation

3.1.1 Murine Bone-Marrow Derived Dendritic Cell Differentiation

Perform all procedures in a laminar flow hood to maintain a sterile environment at all times. All animal procedures must be carried out by adequately trained staff and comply with rules set forth by your designated animal facility. Use BMDC medium, PBS and EDTA solutions, pre-warmed to 37 °C unless otherwise stated.

1. **Day 0:** Sacrifice mice by a method in accordance with your institute's animal regulations.
2. Using sterile scissors, cut away as much fat and muscle as possible from the femur and tibia and place the leg in a 50 mL tube containing flushing medium on ice.
3. Transfer the bones with medium into a petri dish and separate femurs from the tibiae.

4. Fill a 20 mL syringe with flushing medium, attach a 27G needle and flush the bone marrow through the bones into a fresh sterile petri dish (*see* **Note 19**).
5. Break up cell aggregates by passing the cells repeatedly through a 20 mL syringe equipped with a 19G needle.
6. Transfer cells to a 50 mL tube and centrifuge at 300 × *g* for 5 min.
7. Pour off the supernatant and resuspend the cell pellet in 1 mL of ammonium chloride solution to lyse red blood cells. Leave the ammonium chloride solution on the cells for 2 min at RT and then fill the tube with flushing medium (~ 40 mL) and mix by inversion.
8. Centrifuge the tubes at 300 × *g* for 5 min.
9. Pour off the supernatant and resuspend the cell pellet in 10 mL of BMDC medium.
10. Count the cells (*see* **Note 20**) and adjust to the required concentration with BMDC medium (*see* **Note 21**).
11. Differentiate the cells into DCs in T-175 tissue culture flasks: add between 30 and 38 mL of cell suspension at 1×10^6 cells/mL per flask and incubate at 37 °C, 5 % CO_2 for 3 days.
12. **Day 3**: Feed cells by adding 30 mL of pre-warmed BMDC medium, per flask and incubate at 37 °C, 5 % CO_2 for another 3 days.
13. **Day 6**: For each tissue culture flask, fill two 50 mL tubes with 10 mL of pre-warmed BMDC medium.
14. Carefully take the flasks out of the incubator & discard the supernatant by pipetting gently.
15. Add 30 mL of sterile pre-warmed PBS to each flask. Gently agitate the flask by moving contents from side-to-side. PBS should appear cloudy as loosely adherent cells become detached.
16. Gently pipette the PBS solution over the adherent/semi-adherent cells (flushing) approximately 10 times.
17. Transfer the PBS containing cells from the flask to one of the 50 mL tubes containing 10 mL BMDC medium.
18. Add 25 mL of pre-warmed EDTA to each flask and incubate at 37 °C for 10 min.
19. Gently agitate the flask, flush the cells 5×, as in **steps 15** and **16** above, and transfer the cell suspension into the second 50 mL tube containing 10 mL BMDC medium.
20. Centrifuge both 50 mL tubes at 300 × *g* for 5 min, pour off the supernatant, and then pool and resuspend the cell pellets in 10 mL of BMDC medium.

21. Count the cells using a 1:10 dilution (*see* **Note 20**) and adjust the cells to a concentration of 7×10^5 cells/mL (for mouse strains BALB/c, C3H/HeN, or C3H/HeJ) or 4.5×10^5 cells/mL (for mouse strain C57BL/6) using BMDC medium.
22. Continue to differentiate the cells into DCs in a T-175 tissue culture flask by adding between 30 and 38 mL of cell suspension per flask and incubating at 37 °C, 5 % CO_2 for a further 2 days.
23. **Day 8**: Feed cells by adding 30 mL of pre-warmed BMDC medium per flask and incubate at 37 °C, 5 % CO_2 for a further 2 days.
24. **Day 10**: Remove loosely adherent cells by gentle repeat pipetting off the medium and then transfer to a 50 mL tube and centrifuge at $300 \times g$ for 5 min.
25. Pour off the supernatant and pool pellets from each tube into 10 mL of BMDC medium.
26. Count cells using a 1:10 dilution (*see* **Note 20**) and adjust the cells to a concentration of 6.25×10^5 cells/mL if assessing cytokine responses, or $1–2 \times 10^6$ cells/mL if using cells for Western blot or FACS analysis, using BMDC medium.
27. Plate cells in round bottomed 96-well plates for cytokine analysis (200 μL/well) or 12-well/24-well plates (1–2 mL/well) for protein/FACS analysis and incubate at 37 °C, 5 % CO_2 overnight.
28. **Day 11**: Stimulate cells as required (*see* **Notes 22** and **23** and Subheading 3.1.2).

3.1.2 Priming and Stimulation of BMDCs

1. Prime cells using TLR agonists (e.g., LPS at 1 ng/mL final concentration) for at least 2 h prior to incubation with particulate materials.
2. Activate the inflammasome with particulates as outlined in **item 6** of Subheading 2.1.2 at the concentrations indicated.
3. Stimulate cells in triplicate and include triplicate blank wells and all relevant positive and negative controls.
4. After a 24 h incubation period, centrifuge the cells at $300 \times g$ for 5 min and collect the supernatant from the plates for ELISA analysis, according to manufacturer's instructions.

3.1.3 ELISA Analysis of BMDC Supernatants

ELISA should be carried out according to the manufacturer's instructions. Detection of specific cytokines or chemokines requires different concentrations of antibodies, blocking solutions and different incubation times. A general procedure is outlined below:

1. Coat the 96-well plate: dilute capture antibody in coating buffer as per manufacturer instructions and transfer 40–100 μL to each well of a high affinity, protein-binding ELISA plate.

2. Seal plate to prevent evaporation and incubate at 4 °C overnight.
3. Bring the plate to RT, flick off the capture antibody solution and wash plate 3× with ELISA wash buffer (*see* Subheading 2.1.3).
4. Block nonspecific binding sites by adding 200 μL/well of blocking solution to each well.
5. Seal plate and incubate at RT for 1 h.
6. Wash plate 3× with ELISA wash buffer and firmly blot plate against clean paper towels.
7. Dilute standards and samples to desired concentrations in ELISA dilution buffer and add 40–100 μL/well to the plate.
8. Seal the plate and incubate at RT for 2 h or at 4 °C overnight.
9. Wash plate 5× with ELISA wash buffer and firmly blot plate against clean paper towels.
10. Dilute biotin-labelled detection antibody in ELISA dilution buffer, as per manufacturer instructions. Add 40–100 μL/well of diluted antibody to plate.
11. Seal the plate and incubate at RT for 1 h.
12. Wash plate 3× with ELISA wash buffer and firmly blot plate against clean paper towels.
13. Dilute Avidin-HRP conjugate or other enzyme conjugate in ELISA dilution buffer, as per manufacturer instructions. Add 40–100 μL/well onto plate.
14. Seal the plate and incubate at RT for 30–60 min.
15. Wash plate 5× with ELISA wash buffer and firmly blot plate against clean paper towels.
16. Transfer 40–100 μL/well of HRP substrate solution into plate and incubate at RT for 5–30 min, while monitoring color development.
17. Stop color development reaction by adding 20–50 μL/well of stop solution.
18. Read the optical density (OD) for each well with a microplate reader set to 492 nm.

3.1.4 Western Blot Analysis

Western blot analysis should be carried out when BMDCs have been differentiated appropriately and plated on day 10 (*see* Subheading 3.1.1).

1. Plate differentiated BMDCs in a 12-well plate at a density of $0.5–1 \times 10^5$ cells/mL (1 mL/well) in BMDC stimulation medium containing less than 5 % FCS (*see* **Notes 24** and **25**) and incubate at 37 °C, 5 % CO_2 overnight.

2. On the following day, carefully remove and discard 0.5 mL of supernatant from each well. If using 6-well plates, remove 0.75–1.0 mL/well.
3. Prime with TLR ligands (e.g., LPS at final concentration of 1 ng/mL for BMDCs) in 10–20 μL of BMDC stimulation medium containing no more than 5 % of FCS and incubate for 3 h at 37 °C, 5 % CO_2.
4. Add particulates (20–40 μL/well of particulates for a maximum of 250 μL/well) to a microplate at optimized concentrations. Incubate for 3, 6, 18 and 24 h (*see* **Note 26**) for detection of cleaved caspase-1 and/or IL-1β.
5. Collect supernatant using a micropipette with sterile tips into 1.5 mL centrifuge tubes. Spin tubes at 500 × *g* for 2 min to remove any cell debris.
6. Add 10 μL of Strataclean® beads/mL of supernatant collected. Make sure Strataclean® beads are fully dispersed in solution by gently vortexing. The solution color should be white.
7. Vortex tubes for 15 s and let tubes stand for 2–3 min. Repeat three times.
8. Centrifuge tubes in a microcentrifuge at 12,000 × *g* for up to 10 min (or until the beads form a distinct depot at the bottom of the tube).
9. Carefully remove the supernatant. Resuspend the beads in a small volume of 1× Laemmli sample buffer (no more than 50–60 μL if you started with 1 mL of supernatant) and vigorously pipette up and down 6×.
10. Heat tubes at 95 °C for a maximum of 10 min.
11. Spin microcentrifuge tubes at 12,000 × *g* for up to 10 min.
12. Load 20–30 μL/well of each sample and 5–10 μL/well of prestained markers to a 15 % polyacrylamide gel.
13. Run gel at 100–110 V for 1–2 h until the migration front reaches about ¾ of the way down the gel (*see* **Note 27**).
14. Prepare a 20 % MeOH Western blot transfer buffer and proceed to either wet or dry transfer—this is a personal preference (*see* **Note 28**).
15. Once transfer has completed, block the membrane(s) with 5 % nonfat milk in TBST, for 30 min at RT with agitation.
16. Wash membrane 3× for 5 min with TBST at RT with agitation.
17. Incubate PVDF membrane with rabbit anti-mouse caspase-1 primary antibody at 1:1,000 dilution in 5 % nonfat milk in TBST, overnight at 4 °C.
18. Wash membrane 5× for 3 min using TBST with agitation at RT.

19. Incubate membrane with HRP-conjugated anti-Rabbit antibody at 1:10,000 dilution in 5 % nonfat milk in TBST, for 2 h at RT.
20. Wash membrane 6× for 3 min with TBST at RT with agitation.
21. Proceed to ECL reaction using in-house solutions or Millipore ECL detection reagent A + B solution (*see* **item 26** of Subheading 2.1.4). Incubate membrane in solution for 2 min then blot dry.
22. Expose membrane to films in dark room and run through developer machine.
23. If probing for IL-1β: strip membrane for a maximum of 10 min using Western Blot Stripping buffer or as instructed by manufacturer.
24. Wash membrane 3× for 5 min with TBST at RT with agitation.
25. Incubate membrane with goat anti-mouse IL-1β at 1:1,000 dilution in 5 % nonfat milk in TBST, overnight at 4 °C.
26. Wash membrane 5× for 3 min with TBST at RT with agitation.
27. Incubate PVDF membrane with HRP-conjugated anti-Goat at 1:20,000 dilution in 5 % nonfat milk in TBST, for 2 h at RT.
28. Wash membrane 6× for 3 min with TBST at RT with agitation.
29. Proceed to ECL reaction using in-house solutions or Millipore ECL detection reagent A + B solution (*see* **item 26** of Subheading 2.1.4). Incubate membrane in solution for 2 min then blot dry.
30. Expose membrane to films in dark room and run through developer machine (*see* Fig. 2 [10]).

3.1.5 Determination of ASC-Activation Using Confocal Bio-imaging of Live THP-1 Cells

The THP1-ASC-GFP cells represent an ideal in vitro system to study and observe, in real time or by time-lapse imaging, the effect of a broad range of stimuli on the behavior of ASC in live cells and the consequences of ASC oligomerization on cell viability [13].

1. Pellet exponentially growing THP-1–ASC–GFP cells by centrifugation at 300 × *g* for 5 min.
2. Resuspend the cells in fresh DMEM medium containing phorbol 12-myristate-13-acetate (PMA) (0.5 mM) at a density of 5×10^5 cells/mL. Pipette 1 mL of the suspended cells into 35 mm cover glass bottom culture dishes and allow cells to adhere for 2 h.
3. Remove the PMA-containing medium and replace it with fresh DMEM medium without PMA (*see* **Note 29**).
4. If nuclear and/or plasma membrane staining is desired, stain cells with Hoechst 33342 and/or CellMask™ Red plasma membrane stains according to the manufacturer's instructions.

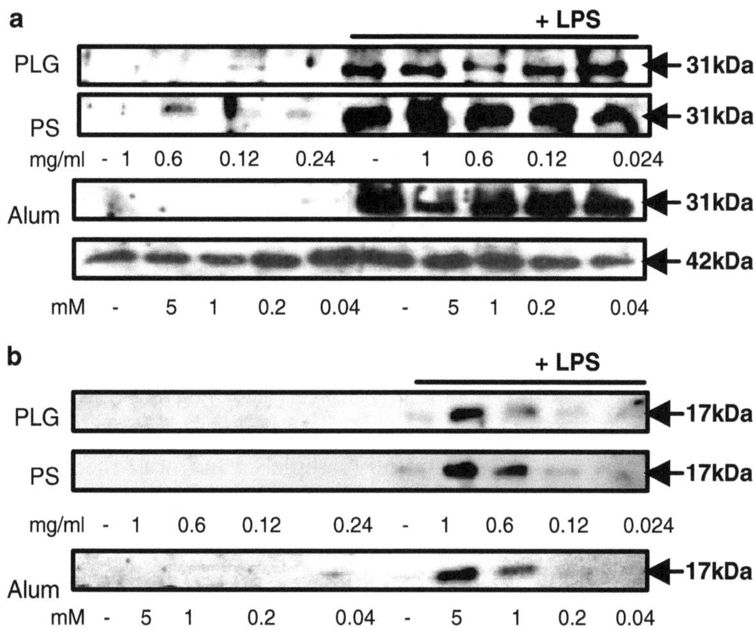

Fig. 2 Western blot analysis. BMDC cultures were incubated with particles alone (poly(lactide-co-glycolide) (PLG), polystyrene (PS) microparticles, or Alhydrogel®), LPS alone, or LPS + particles for period of 24 h. (**a**) Pro-IL-1β (31 kDa) and β-actin (42 kDa) in cell lysates. (**b**) Mature IL-1β (17 kDa) in supernatants

5. Place the culture dish in the microscope sample chamber maintained at 37–38 °C and 5 % CO_2 (*see* **Note 30**).
6. Stimulate the cells with LPS (1–5 μg/mL) or other pro-inflammatory stimuli. Ten minutes after LPS stimulation, start recording images from the GFP, Hoechst, and other fluorescence markers simultaneously every 10 s for 30–60 min (*see* **Note 31**).
7. Add particulate stimuli (e.g., MSU, alum salts) to petri dish at concentrations deemed optimal and continue recording (*see* **Note 32**).

3.2 Measuring Cytotoxicity

3.2.1 BMDC Cultures

1. Grow and differentiate BMDC cultures until day 10, as described in Subheading 3.1.1.
2. On day 10, take out T-175 tissue culture flask and verify cell viability and morphology under a light microscope (*see* **Note 33**).
3. Gently pipette the BMDCs cultures up and down to detach cells, as described in Subheading 3.1.1. BMDCs should detach from the bottom of the flask.
4. Transfer the cell suspension into 50 mL tubes and centrifuge at 300 × *g* for 5 min.
5. Discard supernatant and resuspend cell pellet in 5 mL of BMDC medium and count the cells (*see* **Note 20**).

6. Plate 5–10 × 10^3 cells per well in a 96-well plate (or 5–10 × 10^4 cells per well in a 24-well plate) and incubate overnight at 37 °C, 5 % CO_2 to allow cells to adhere.
7. Stimulate BMDCs cultures with particulates at various concentrations in order to assess dose–response effects. Use untreated BMDCs as a negative control and BMDCs stimulated with apoptosis- or necrosis-inducing agents as positive controls (*see* **Note 34**).
8. Incubate cells with particulates (*see* **Notes 35** and **36**) for a predefined period of time, e.g., 3, 6, and 24 h at 37 °C (*see* **Note 37**).

3.2.2 AnnexinV-PI Binding Assay

1. Centrifuge treated BMDC single cell suspension at 300 × *g* for 5 min, discard supernatants and resuspend in 200 μL binding buffer/sample (at no more than 2 × 10^6 cells/mL).
2. Incubate BMDC samples with 1 μg/mL Annexin V-FITC (or at the concentration deemed optimal for your setup) for 5 min.
3. Centrifuge at 500 × *g* for 5 min, discard supernatant and resuspend in 200 μL of binding buffer.
4. Transfer each sample into appropriately labelled FACS tubes.
5. Add PI directly into FACS tubes at a final concentration of 1 μg/mL, just prior to the acquisition of samples on the flow cytometer (*see* **Note 38**).
6. Analyze samples by flow cytometry within 1 h (*see* **Note 39**).

3.2.3 LIVE/DEAD® Aqua Staining

1. Bring one vial of LIVE/DEAD® aqua fluorescent reactive dye (*see* **Note 40**) and one vial of anhydrous DMSO (*see* **Note 41**) to RT.
2. Add 50 μL of anhydrous DMSO to one vial of reactive dye, as per manufacturer instructions (*see* **Note 42**). Mix well and visually confirm that the dye has completely dissolved.
3. At the end of the BMDC stimulation period with particulate compounds, centrifuge microplate at 300 × *g* for 5 min.
4. Discard the supernatant and resuspend in 200 μL/well of PBS for 96-well plate.
5. Count BMDCs and adjust density with PBS such that cell density is no more than 1 × 10^6 cells/mL, as per manufacturer instructions.
6. Add 1 μL of the reconstituted fluorescent reactive dye to 1 mL of the BMDC suspension and mix well.
7. Incubate on ice for 15–20 min, protected from light.
8. Centrifuge the cells at 300 × *g* for 5 min, discard supernatants and resuspend in 200 μL/well of PBS.
9. If fixation is required, carefully remove supernatant and add 100–200 μL/well of cell fixative (*see* **Notes 43** and **44**).
10. Incubate at RT for 10 min.

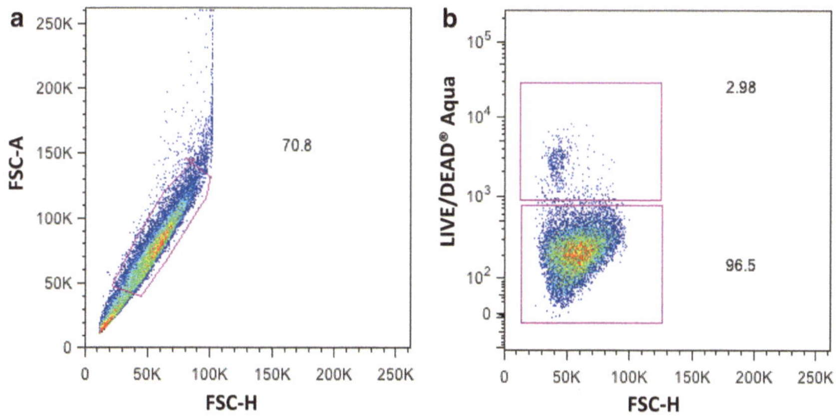

Fig. 3 Common gating strategy to remove doublets and dead cells in the cell analysis. (**a**) The first histogram uses the measurements of forward scatter area (FSC-A) and forward scatter height (FSC-H) to determine the single cells (*pink polygon*) from the doublets. (**b**) Following the staining reaction with LIVE/DEAD aqua, the cells were fixed in BD Cytofix/Cytoperm™ and analyzed by flow cytometry. The live cell population (*lower panel*, 96.5 %) is easily distinguished from the killed population (*upper panel*, 2.98 %)

11. Centrifuge the cells at 300 × *g* for 5 min, discard supernatant and wash cells once with 200 μL/well of FACS staining buffer.
12. Analyze samples by flow cytometry within 48 h using an analyzer equipped with a violet laser (*see* Figs. 3 and 4, and **Note 45**).

4 Notes

1. Annexin V-PI staining is a convenient way to discriminate early apoptosis from late apoptosis and necrosis. Early apoptotic cells express phosphatidylserine (PS) on the outer leaflet of the plasma membrane and PS can thus be stained with labelled annexin V. In contrast, late apoptotic cells and necrotic cells lose their cell membrane integrity and are permeable to vital dyes, such as PI.
2. The compound 7-amino-actinomycin (or 7-ADD) can be used as a DNA marker—it forms stable complexes with double strand DNA—in lieu of PI, at a concentration of 50 mg/mL, i.e., 20 μL/100 μL sample, e.g., use 1×10^6 cells in a 100 μL experimental sample. The maximum absorption of the 7-AAD/DNA complex is compatible with the blue excitation wavelength of 488 nm for cytometers fitted with an argon laser. The peak fluorescence emission in the deep red band (635–675 nm) of the 7-AAD/DNA complex permits optimal use of this probe when it is combined with Fluorescein Isothiocyanate conjugated antibodies (FITC) and R Phycoerythrin (PE). Indeed, in contrast to PI, the 7-AAD/DNA complex has a reduced overlapping emission spectrum with FITC and PE.
3. Regarding the performance of the LIVE/DEAD® aqua stain, the manufacturer states that the difference in intensity between live and dead cell population is greater than 50-fold.

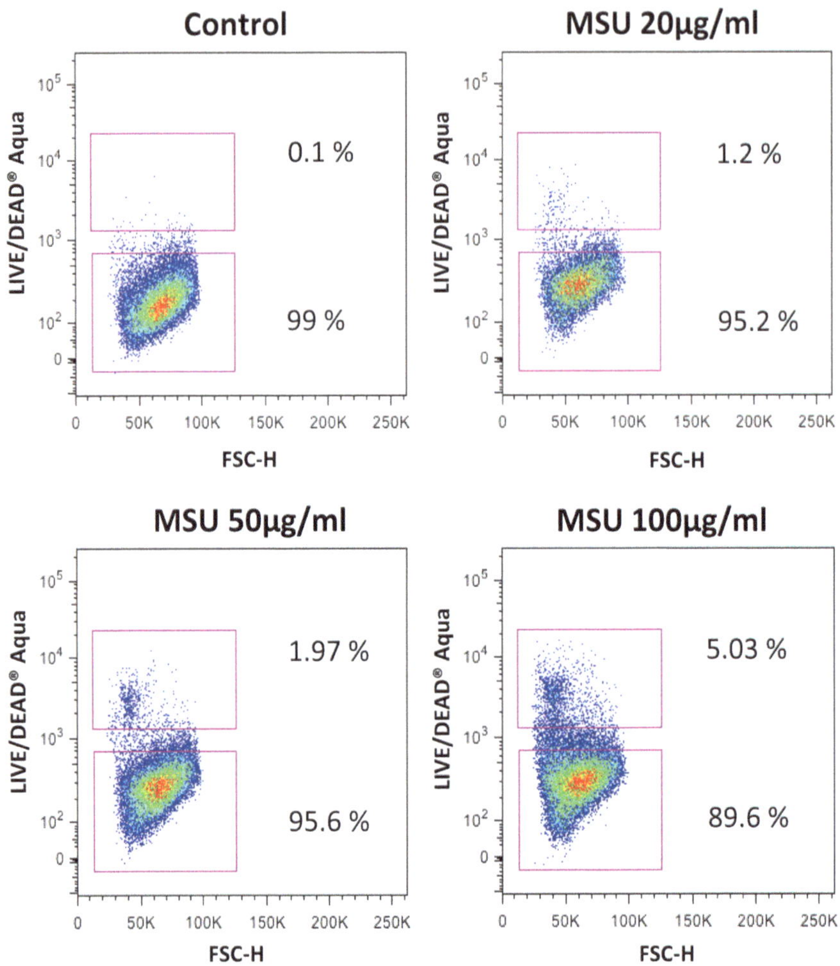

Fig. 4 Dose–response analysis of MSU-induced cell toxicity in BMDC cultures (C57BL/6) after 6 h incubation. MSU was used at concentrations ranging from 20 to 100 μg/mL

4. Depending on the age and strain of mice, and on the efficiency of the procedure, one mouse typically generates up to $20–40 \times 10^6$ BMDCs.
5. Alternatively, a commercially available solution of red blood cell (RBC) lysis buffer can also be used.
6. As a more economical alternative to the use of commercial recombinant mouse GM-CSF, prepare BMDC differentiation medium supplemented with 20 % conditioned medium from mouse GM-CSF producing J558 cells [14].
7. We recommend that other cytokines, e.g., IL-6, TNF-α, and IL-12 concentrations be measured to ensure that NLRP3 inflammasome activation is particle-driven and specific, and that chemical inhibitors are not modulating inflammasome-independent cytokines.

8. The apparent molecular masses of the pro- and cleaved forms of IL-1β are approximately 31 kDa and 17 kDa respectively, in SDS-PAGE under reducing conditions, while the caspase-1 precursor and cleaved forms have apparent molecular weights of approximately 45 kDa and 10 kDa, respectively. In order to obtain optimal separation between these two proteins and their cleaved form, we found that 15 % polyacrylamide gels were optimal.
9. The ability of membranes to bind proteins varies greatly depending on the amounts of protein present in the samples, the incubation times and the potential loss of proteins following incubation with different detergents or blocking agents. We compared the quality of the immunoblots using PVDF vs. nitrocellulose membrane using the protocol described herein and we consistently found that PVDF membrane performed better than nitrocellulose, particularly for the immunodetection of cleaved caspase-1. Note that PVDF membranes must be hydrated in 100 % methanol and then rinsed with deionized water before being placed in the transfer buffer to equilibrate.
10. We recommend the use of a low range prestained protein ladder, e.g., 3.4–100 kDa. It will serve the double purpose of confirming the apparent molecular weight of your protein bands and provide an indication on the efficiency of your protein transfer from the gel onto the membrane.
11. Our lab determined that the Restore™ (or Restore™ PLUS) Western blot stripping buffer from Thermo Scientific performed particularly well.
12. An alternative to ECL detection is the use of infrared detection (Odyssey®).
13. This chapter details materials and methods for determining the cytotoxic effects of particulate materials on BMDC cultures, but a similar approach is applicable to other antigen presenting cells, such as macrophages. Other cell types, doses and concentrations will require prior optimization. For titration procedures, detailed protocols have been described elsewhere [15].
14. Depending on the amount of particulate materials available for use, a plate format or 2 mL reaction tubes may be used.
15. A recent study described a detailed step-by-step procedure to express and purify milligram amounts of "home-made" annexin V within 3 days [16]. The method explains the expression of recombinant annexin V in competent DH5α *E. coli* bacteria, its purification using NI-NTA agarose, and its conjugation to the FITC fluorochrome. More specifically, human annexin V was cloned into the bacterial expression vector pProEx.Htb which encodes ampicillin resistance and incorporates an

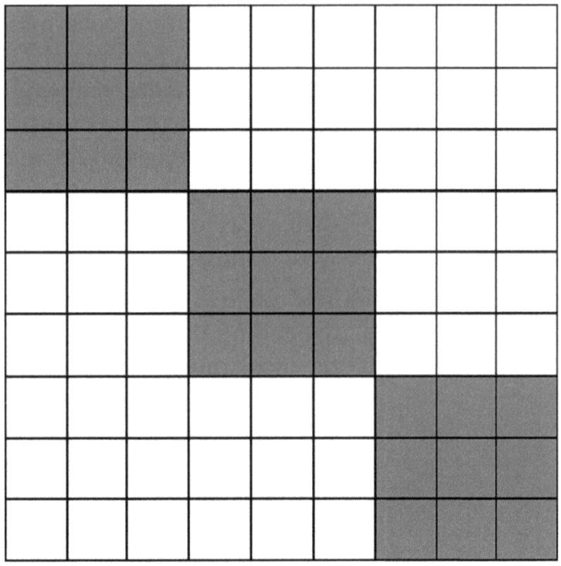

Fig. 5 Hemocytometer chamber and cell counting strategy

N-terminal poly-histidine tag. The resulting plasmid (i.e., pProEx.Htb.annexin V) is made freely available upon request to investigators working in nonprofit organizations.

16. General considerations for a successful Annexin-V binding assay: cells should *not* be fixed or permeabilized (e.g., paraformaldehyde or methanol) and similarly, trypsinization or scraping to detach cells should be carefully considered as this may damage the plasma membrane, resulting in false positives [16].
17. A recent study [16] determined that calcium was required at a concentration of 1–3 mM for optimal binding of annexin V to PS. For most cell types, a concentration of 1.8 mM calcium is ideal.
18. Cell fixative: either BD Cytofix/Cytoperm™ solution or a solution of 0.4 % paraformaldehyde may be used.
19. When flushing the femur and tibiae through with medium using a 27G needle, the flushing liquid coming out should be a red color and if flushed through correctly the bone should become clear/white when all the cells are flushed out. To ensure all cells are removed, flush each bone through from both ends.
20. A 1:10 dilution is usually adequate for cell counting, i.e., 90 μL of trypan blue solution and 10 μL of cells, mix well and transfer to a hemocytometer slide. Count the cells as follows: count three larger squares (of 9 smaller squares each, *see* Fig. 5), get the average of the three and calculate the number of cells per mL using the following formula:

$$\text{Cells/mL} = \text{average cell count} \times \text{dilution factor } (10) \times 10^4$$

21. Our lab has determined that different mouse strains required different seeding density on day 6 in order to reach optimal growth and differentiation on day 10.
22. Investigators may wish to use chemical inhibitors such as YVAD-fmk (caspase-1 inhibitor), bafilomycin A (inhibitor of the vacuolar H+ ATPase), or CA-074-Me (cathepsin B inhibitor) in order to gain further insight into the mechanisms of inflammasome activation [10]. The materials in Subheading 2.1.2 should also be used when planning to assess the effects of chemical inhibitors. They may be used in a similar manner to that described for priming and stimulating BMDCs (Subheading 3.1.2) and may be added before, at the same time, or after BMDCs have been primed. Advised concentrations of inhibitors are final concentrations and should be adjusted depending on the volume of cells in the well and the volume of the inhibitor to be added. Concentrations should be optimized in preliminary experiments or should be sourced from the literature. Note that chemical inhibitors could also affect priming (e.g., induction of NF-κB), thus, caution should be taken and controls included, ensuring that the inhibitor effect can be attributed to modulation of inflammasome activation rather than pro-IL-1β or IL-18 induction.
23. When priming and/or stimulating cells, note that the concentrations advised are for a final concentration in a well, i.e., 40 mg/mL silica final concentration. For example in a 96-well plate if there are 200 μL cells already in the well and the advised amount of 25 μL of the stimulant made up in priming medium is added, the concentration in the well will be diluted 1 in 9. Therefore stock solutions should be made up to a concentration nine times greater than the final concentration and 25 μL added per well in a 96-well plate.
24. We recommended using no more than 5 % FCS, to avoid the possibility of protein overloading which could negatively affect the quality of protein band signals. If difficulties are encountered (e.g., poor band resolution or slow migration during electrophoresis), decrease the concentration of FCS to 2 % and incubate for a maximum of 6 h. A further decrease in FCS concentration will affect the responsiveness of the BMDCs to TLR stimulation and/or particulate materials, and thus the quality of the signal.
25. We recommend running each experimental condition in triplicate wells so supernatants can be pooled, thus increasing signal intensity.
26. The optimal incubation period will be largely dependent on the characteristics of your particulate material. Longer incubation periods may be required depending on the particle–cell

ratio and the quality of the cell preparation, e.g., cell viability and maturation. Using the protocol described herein, our lab has determined that an incubation period of 3–6 h was optimal for the detection of caspase-1 and IL-1β in cell supernatants upon stimulation with particulates such as carbon nanotubes, silicon nanoparticles, alum, MSU, and polystyrene beads.

27. The use of prestained molecular weight markers helps to ensure that the caspase-1 p10 protein band remains within the gel; however, its apparent molecular mass may be lower, e.g., 7 kDa [12]. Thus, we recommend not allowing the migration front to go beyond ¾ way down the gel, as the caspase-1 (p10) band signal may be lost.
28. For wet transfer, proceed to transfer at 25 V (maximum 400 mA) and at 4 °C with stirring for at least 2 h. For dry transfer, a transfer at 15 V for 30–60 min should be sufficient.
29. At this stage, the cells are ready to be treated with particulate materials and/or pro-inflammatory agents. Cells are usually left in fresh medium without PMA overnight, but shorter times can be used.
30. We use an Olympus FV1000 confocal microscope system controlled by Olympus FLUOVIEW® viewer software and equipped with a temperature- and CO_2-regulated chamber, but comparable microscopes can be used. It is critical that the temperature of the sample be maintained between 37 and 38 °C and CO_2 be maintained at 5 %. Temperatures below 37 °C may reduce ASC-oligomers formation.
31. For example, use the following settings: GFP is excited with a 488 nm argon laser and its 505- to 545-nm emitted light is collected in a photomultiplier tube (PMT). Hoechst is excited with a 405 nm HeNe laser and its 420- to 480-nm emitted light is collected in a second PMT. Transmitted laser light is collected to produce a differential interference microscopic image. The eight-bit images are converted to a RGB format, the median is filtered to eliminate noise, montaged or overlayed, and can be converted into movie files at five frames per second using software such as Metamorph 7.3 (Molecular Devices Corp.).
32. In unstimulated cells, ASC–GFP should be distributed evenly in the cytoplasm and nucleus. However, ASC begins to oligomerize and may form single clusters as early as 10 min after stimulation with crude LPS or other stimuli [13]. ASC oligomers (observed as bright green specks) may not form simultaneously in all cells and some cells may not form any oligomers even after prolonged stimulation. Usually one ASC oligomer only will be formed in each cell, and this will precede morphological changes characteristic of pyroptosis. See supplementary movies

in reference [13]. ASC oligomer-induced pyroptosis can thus be seen within a period of 1–3 h and be simultaneously monitored by the release of intracellular LDH [13].

33. Cells should appear large and irregularly shaped.
34. Solutions of cycloheximide (protein synthesis inhibitor) and actinomycin D (RNA synthesis inhibitor) or a solution of staurosporine (a wide spectrum apoptosis inducer) can be used as positive controls. Concentrations and incubation times for necrosis/apoptosis induction may vary depending on the cell type used and the specific experimental set up, and thus, should be optimized in preliminary experiments.
35. It is critical to bring potential endotoxin contamination to low pg/mL values. We have observed that LPS concentrations as low as 10 and 20 pg/mL were able to induce secretion of IL-6 and TNF-α respectively, in BMDC differentiated cultures [17].
36. Further assessment of the innate immunomodulatory properties of particulate adjuvants or materials may be carried out in vivo using the mouse peritonitis model: mice can be challenged intraperitoneally with particulate solutions (50–200 μL/mouse at a range of concentrations, e.g., 100–1,000 μg of particles per mouse) for a period of 1–24 h in order to characterize inflammatory cell recruitment patterns. Peritoneal lavages are collected using 3–5 mL of sterile PBS solution. Mice challenged with PBS serve as negative controls. Peritoneal cells collected at different time points can then be stained for extracellular markers and analyzed by multicolor flow cytometry. A classic observation is the recruitment of inflammatory cells such as granulocyte and monocyte populations while resident macrophages are depleted. Further details can be found elsewhere [5, 10].
37. In preliminary experiments, we recommend incubation of the particulate material for a period of 24 h and determine if concentrations of IL-1β are significantly different from negative control wells. A dose–response may be evident between 3 and 24 h, i.e., the longer the incubation period, the higher the concentrations of IL-1β responses in the supernatant. However, this is not always the case, as potent stimuli cytokine secretion may be maximal after only a few hours.
38. Nonviable cells take up PI and its fluorescence can be detected on the FL2 channel (on most flow cytometers) while apoptotic cells display phosphatidylserine on their surface, which binds annexin V such that annexin V- FITC be detected on the FL1 channel (on most flow cytometers).
39. When carrying out annexin V-PI assay, samples should be analyzed within 1 h of completion of cell staining, due to toxicity issues and the risk of generating false positives. We suggest the

following gating strategy: exclude small cell particles or dead cell debris, which are recognized as a tail in the lower part of the dot plot. Then, set up the flow cytometer such that the distribution of the annexin V-FITC negative population is in the first two decades of the FL1 parameter and the distribution of the PI negative population is in the first two decades of the FL2 parameter. BMDCs treated with necrosis or apoptosis-inducing agents are used to set up single and double positive populations, thus gating at least 95 % of dead (early and late apoptotic or secondary necrotic) cells. Untreated BMDCs are used to set apart 95 % of viable cells. When using multicolor flow cytometry, e.g., FITC–annexin V and PI staining, it is important to compensate for spectral overlap between the different fluorochromes used. Refer to flow cytometer manual for specific instructions regarding compensation settings.

40. Other dyes besides LIVE/DEAD® Aqua are available. The single-color LIVE/DEAD® Fixable Dead cell stain kits from Invitrogen are identical except for the fluorescent color of the included dye, i.e., blue, violet, aqua, yellow, green, red, far red, or near-infra red (IR). Each of them has their specific excitation and emission wavelengths. For further details, see: http://www.invitrogen.com/
41. DMSO is hazardous; avoid contact with skin and eyes and do not swallow. Handle reagents containing DMSO using equipment and practices appropriate for the hazards posed by such materials.
42. This solution will be further diluted 1:1,000 in cell suspension.
43. Permeabilization and fixation reagents such as BD Cytofix/Cytoperm (BD Pharmingen) or 0.5 % PFA will not change the stability of the LIVE/DEAD® aqua dye nor will the dye leak out of the cells.
44. Staining for cell surface markers should be performed prior to fixation of the samples but after staining with LIVE/DEAD® aqua stain.
45. The Aqua fluorescent reactive dye (Invitrogen) is efficiently excited with ~405 nm light (or UV light) and has fluorescence emission monitored at ~525 nm.

References

1. Couillin I, Petrilli V, Martinon F (2011) The inflammasomes. Progress in inflammation research. Springer, Basel
2. Valavanidis A, Fiotakis K, Vlachogianni T (2008) Airborne particulate matter and human health: toxicological assessment and importance of size and composition of particles for oxidative damage and carcinogenic mechanisms. J Environ Sci Health C Environ Carcinog Ecotoxicol Rev 26(4):339–362

3. Strowig T, Henao-Mejia J, Elinav E, Flavell R (2012) Inflammasomes in health and disease. Nature 481(7381):278–286
4. Hornung V, Bauernfeind F, Halle A, Samstad EO, Kono H, Rock KL (2008) Silica crystals and aluminum salts activate the NALP3 inflammasome through phagosomal destabilization. Nat Immunol 9(8):847–856
5. Martinon F, Petrilli V, Mayor A, Tardivel A, Tschopp J (2006) Gout-associated uric acid crystals activate the NALP3 inflammasome. Nature 440:237–241
6. Dostert C, Pétrilli V, Van Bruggen R, Steele C, Mossman BT, Tschopp J (2008) Innate immune activation through Nalp3 inflammasome sensing of asbestos and silica. Science 320(5876):674–677
7. Li H, Willingham SB, Ting JP, Re F (2008) Cutting edge: inflammasome activation by alum and alum's adjuvant effect are mediated by NLRP3. J Immunol 181(1):17–21
8. Mori A, Oleszycka E, Sharp FA, Coleman M, Ozasa Y, Singh M, O'Hagan DT, Tajber L, Corrigan OI, McNeela EA, Lavelle EC (2012) The vaccine adjuvant alum inhibits IL-12 by promoting PI3 kinase signaling while chitosan does not inhibit IL-12 and enhances Th1 and Th17 responses. Eur J Immunol. doi: 10.1002/eji.201242372
9. Bueter CL, Lee CK, Rathinam VA, Healy GJ, Taron CH, Specht CA, Levitz SM (2011) Chitosan but not chitin activates the inflammasome by a mechanism dependent upon phagocytosis. J Biol Chem 286(41):35447–35455
10. Sharp FA, Ruane D, Claass B, Creagh E, Harris J, Malyala P, Singh M, O'Hagan DT, Pétrilli V, Tschopp J, O'Neill LAJ, Lavelle EC (2009) Uptake of particulate vaccine adjuvants by dendritic cells activates the NALP3 inflammasome. Proc Natl Acad Sci USA 106(3):870–875
11. Demento SL, Eisenbarth SC, Foellmer HG, Platt C, Caplan MJ, Mark Saltzman W, Mellman I, Ledizet M, Fikrig E, Flavell RA, Fahmy TM (2009) Inflammasome-activating nanoparticles as modular systems for optimizing vaccine efficacy. Vaccine 27(23):3013–3021
12. Doyle SL, Campbell M, Ozaki E, Salomon RG, Mori A, Kenna PF, Farrar GJ, Kiang AS, Humphries MM, Lavelle EC, O'Neill LA, Hollyfield JG, Humphries P (2012) NLRP3 has a protective role in age-related macular degeneration through the induction of IL-18 by drusen components. Nat Med 18(5):791–798
13. Fernandes-Alnemri T, Wu J, Yu JW, Datta P, Miller B, Jankowski W, Rosenberg S, Zhang J, Alnemri ES (2007) The pyroptosome: a supramolecular assembly of ASC dimers mediating inflammatory cell death via caspase-1 activation. Cell Death Differ 14(9):1590–1604
14. de Brito C, Tomkowiak M, Ghittoni R, Caux C, Leverrier Y, Marvel J (2011) CpG promotes cross-presentation of dead cell-associated antigens by pre-CD8α+ dendritic cells. J Immunol 186(3):1503–1511
15. Perfetto SP, Chattopadhyay PK, Lamoreaux L, Nguyen R, Ambrozak D, Koup RA, Roederer M (2010) Amine-reactive dyes for dead cell discrimination in fixed samples. Curr Protoc Cytom Chapter: Unit–9.34
16. Logue SE, Elgendy M, Martin S (2009) Expression, purification and use of recombinant annexin V for the detection of apoptotic cells. Nat Protoc 4(9):1383–1395
17. Tynan GA, McNaughton A, Jarnicki A, Tsuji T, Lavelle EC (2012) Polymyxin B inadequately quenches the effects of contaminating lipopolysaccharide on murine dendritic cells. PLoS One 7(5):e37261

Website Reference

Invitrogen web site of LIVE/DEAD® fixable dyes: http://www.invitrogen.com/site/us/en/home/Global/invitrogen-search-results.html?searchTerm=live+%2F+dead+fixable&searchRows=15&searchTypes=meta.collection%3Acmgtbuyproduct%3A&zeroCategories=meta.collection%3Amrdb%3A

Chapter 6

Measuring Inflammasome Activation in Response to Bacterial Infection

Petr Broz and Denise M. Monack

Abstract

Inflammasomes are multi-protein signaling platforms assembled in response to viral and bacterial pathogens as well as endogenous danger signals. Inflammasomes serve as activation platforms for the mammalian cysteine protease caspase-1, a central mediator of innate immunity. The hallmarks of inflammasome activation are the processing of caspase-1, the maturation and release of interleukin-1β (IL-1β) and the induction of pyroptosis, a lytic inflammatory cell death. This protocol describes methods for studying inflammasome activation in response to bacterial pathogens in bone-marrow derived murine macrophages (BMDMs). In particular, we outline the protocols to measure cytokine maturation by ELISA and pyroptosis by the release of Lactate Dehydrogenase (LDH). In addition, we describe methods to visualize endogenous ASC specks or foci in infected cells and to study the release of processed caspase-1, caspase-11 and mature cytokines into the cell supernatant by Western blotting. General considerations are discussed to design and optimize the infection protocol for the study of inflammasome activation by other bacterial pathogens.

Key words Inflammasome, Caspase-1, Caspase-11, Pyroptosis, Interleukin-1, LDH release, NLRs, Bacterial infections, ASC speck

1 Introduction

The innate immune system is the first line of defense against microbial infection and is activated by the engagement of germline-encoded pattern-recognition receptors (PRRs). These receptors recognize a variety of unique bacterial molecules as well as endogenous danger signals, and initiate appropriate inflammatory responses. A key event in the inflammatory pathway is the activation of the inflammasome, a multiprotein platform that activates caspase-1 [1]. Activated caspase-1 promotes the maturation and secretion of pro-inflammatory cytokines interleukin-1β (IL-1β) and IL-18 [2, 3] and the release of proteins involved in tissue repair and cytoprotection [4]. In addition, active caspase-1 induces a pro-inflammatory form of cell death called pyroptosis [3, 5].

Christine M. De Nardo and Eicke Latz (eds.), *The Inflammasome: Methods and Protocols*, Methods in Molecular Biology, vol. 1040, DOI 10.1007/978-1-62703-523-1_6,

Inflammasomes typically consist of three main components: (1) a cytoplasmic sensor/receptor, (2) caspase-1, and (3) in most cases, the adaptor protein ASC. Several types of inflammasomes have been identified and are named after the sensor that initiates the assembly of the complex. Thus far, only members of the NLR and the PYHIN protein families were shown to initiate inflammasome activation. These proteins function as specific sensors for various pathogens and endogenous danger signals. For example, Nlrp1b is required for caspase-1 activation in response to anthrax lethal toxin [6]. NLRP3 responds to a large variety of structurally and chemically different molecules, but the molecular mechanism linking these molecules to NLRP3 activation remains poorly understood [7]. Several viral and bacterial pathogens are known to activate the NLRP3 inflammasome [8]. NLRC4 activates caspase-1 after infection with bacteria expressing flagella and/or Type 3 secretion systems (T3SS) (such as *Salmonella* spp., *Pseudomonas aeruginosa*, *Listeria monocytogenes*, and *Legionella pneumophila* and others) [9–13]. In mice Nlrc4 appears to specifically detect flagellin with the help of Naip5/6 and the T3SS rod subunits through Naip2, while human NAIP detects the T3SS needle subunit [14–16]. Finally, AIM2 recognizes the presence of double stranded DNA in the cytoplasm and is activated during infections with certain DNA viruses and the cytosolic bacterial pathogens *Francisella novicida* and *L. monocytogenes* [17–24].

An important component of most inflammasomes is the bipartite adaptor protein ASC, which is required by many NLRs and AIM2 to recruit pro-caspase-1 into the inflammasome complex. ASC stands for "apoptosis-associated speck-like protein containing a CARD," which describes its ability to form a large, macromolecular speck in the cytoplasm [25]. ASC specks, also known as ASC foci or pyroptosomes, form rapidly in response to pathogens [19, 26] or when ASC is over-expressed in cells [27, 28]. Visualization of endogenous ASC foci showed that the structures are 1–2 μm in diameter, serve as an activation platform for pro-caspase-1 [26], and are necessary for efficient pro-IL-1β maturation [29]. In addition, it has been shown that ASC foci co-localize with DNA and AIM2 during *F. novicida* infections (*see* Fig. 1) [19]. Finally, since ASC oligomerization does not require caspase-1 activity, the formation of endogenous ASC foci can be used as a measure for inflammasome complex formation, even in the absence of caspase-1 [26].

Following its activation by dimerization and proteolytic cleavage in the inflammasome, the active caspase-1 initiates several effector mechanisms [3]. The most prominent is the proteolytic maturation and release of the proinflammatory cytokine IL-1β. The release of mature IL-1β is closely controlled by a two-step activation mechanism. The first step (Signal 1) is the induction of pro-IL-1β expression by priming signals, such as the activation of

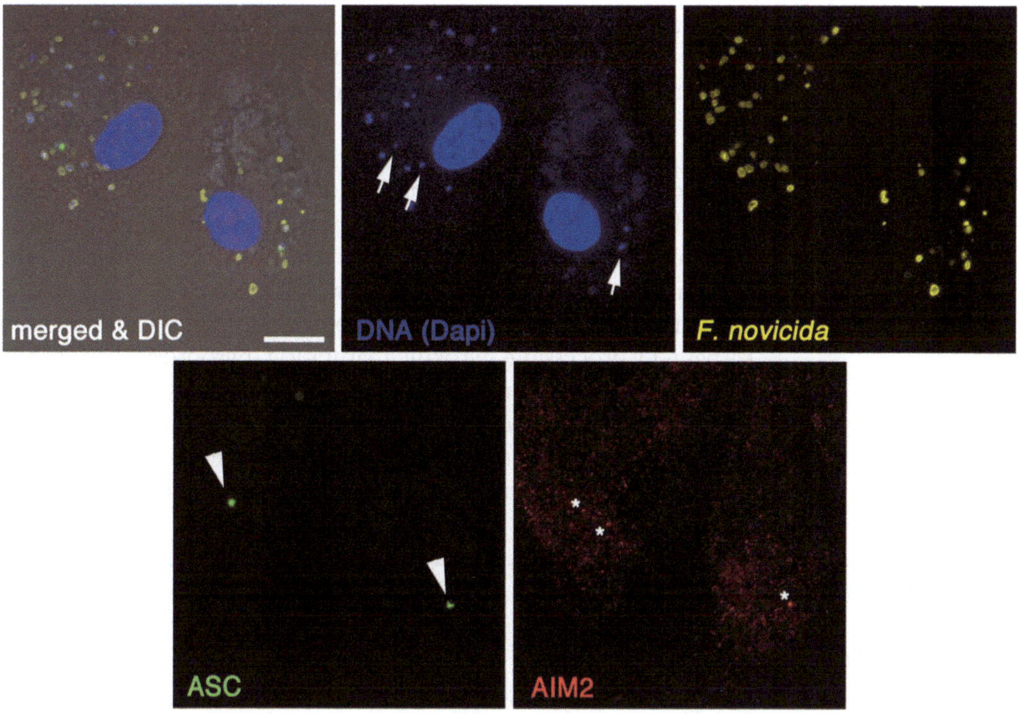

Fig. 1 ASC oligomers (ASC foci or specks) are formed in infected cells. Immunofluorescence staining of wild-type BMDMs infected with *F. novicida*. Cells were fixed and labeled with DAPI (DNA, *blue*), anti-*Francisella* antibodies (*yellow*), anti-ASC antibodies (*green*), and anti-AIM2 antibodies (*red*). *Arrowheads* point out ASC foci, *stars* point to AIM2 speckles, and *arrows* indicate DNA released by lysed bacteria. Scale bar is 10 μm

membrane bound PRRs and the subsequent transcriptional response. The second step (Signal 2) is the activation of the inflammasome itself. However, caspase-1 has also been shown to process and release other cytokines that do not need priming (e.g., IL-18). The exact mechanism by which caspase-1 mediates cytokine release is unknown, but seems to involve a yet uncharacterized secretion pathway, that results in the simultaneous release of processed caspase-1 itself. The second caspase-1 effector mechanism is the induction of a lytic cell death, termed pyroptosis [5], which is thought to remove the replicative niche of intracellular pathogens and reexpose them to extracellular immune responses [30].

Recently, a noncanonical inflammasome was described that activates caspase-11 instead of caspase-1 in murine macrophages [31]. Activation of the noncanonical inflammasome has been linked to TRIF-dependent production of type-I-interferon, and is initiated in response to a variety of Gram-negative bacteria [32, 33]. Similarly to caspase-1 activation, activation of caspase-11 initiates cell lysis and cytokine maturation (in conjunction with NLRP3 and caspase-1), and can thus be analyzed using the same methods that are used to study canonical inflammasomes.

Although some inflammasomes (e.g., NLRP3) detect danger signals, most are dedicated to the recognition of viral or bacterial pathogens. Microbial infections have therefore become an important tool in inflammasome study, since they represent relevant model systems for inflammasome activation, which cannot be achieved using artificial stimuli, such as transfection of DNA. In addition, most pathogens activate several pattern recognition receptors during the course of the infection, thus providing the possibility to study the complex network of innate immune signaling pathways. Here we describe protocols to study caspase-1 and caspase-11 processing, cytokine maturation, pyroptosis, and formation of ASC foci in primary murine BMDMs in response to infections with *F. novicida*, which activates the AIM2 inflammasome. We further discuss considerations to adapt these protocols to study inflammasome activation in response to other bacterial pathogens.

2 Materials

2.1 Differentiation and Long-Term Storage of Murine BMDMs

1. Mice.
2. Autoclaved scissors and forceps.
3. Flushing medium: DMEM high-glucose, unsupplemented.
4. 10 mL syringes.
5. 26Gx1/2 needles.
6. Sterile, non-tissue culture treated 10 cm petri dishes.
7. 70 % Ethanol.
8. Paper towels.
9. Ice.
10. 2 beakers (200–300 mL).
11. Sterile 50 mL tubes.
12. Sterile 5, 10, and 25 mL pipettes.
13. Hemocytometer.
14. Macrophage medium: DMEM supplemented with 10 % FCS, 100 U/mL Pen/Strep, 10 mM HEPES, and 10 % L929-conditioned medium.
15. Centrifuge with rotor fitting 50 mL tubes.
16. Cell culture facility and equipment including 37 °C, 5 % CO_2, water-jacketed cell culture incubator and laminar flow hood.
17. Phosphate buffered Saline (PBS), tissue culture grade.
18. Ice-cold Fetal Calf Serum (FCS).
19. Ice-cold FCS with 20 % DMSO (Dimethyl sulfoxide, tissue culture grade).
20. Cryo vials (1–1.5 mL volume).
21. Isopropanol freezing chamber.

2.2 Thawing and Propagating BMDMs

1. Cryo vials containing 1×10^8 BMDMs/mL.
2. Dry Ice.
3. 37 °C water bath.
4. 70 % Ethanol.
5. Sterile 50 mL tubes.
6. Sterile 5, 10, and 25 mL pipettes.
7. Sterile, non-tissue culture treated 10 cm petri dishes.
8. Macrophage medium.
9. PBS, tissue culture grade.
10. Ice.
11. Centrifuge with rotor fitting 50 mL tubes.
12. Cell culture facility and equipment including 37 °C, 5 % CO_2, water-jacketed cell culture incubator and laminar flow hood.

2.3 Harvesting and Pre-stimulating of BMDMs for an Infection

1. Sterile 50 mL tubes.
2. Sterile 5, 10 and 25 mL pipettes.
3. Sterile, non-tissue culture treated 10 cm petri dishes.
4. Macrophage medium.
5. PBS, tissue culture grade.
6. Ice.
7. Centrifuge with rotor fitting 50 mL tubes.
8. Cell culture facility and equipment including 37 °C, 5 % CO_2, water-jacketed cell culture incubator and laminar flow hood.
9. 1 mg/mL of ultrapure *E. coli* LPS in PBS. Store at −20 °C.

2.4 Measurement of Pyroptosis and Cytokine Maturation

1. Tissue-culture treated flat bottom 96-well plates.
2. Macrophage medium.
3. 12-Tip multichannel pipette, 50–200 μL volume per tip.
4. Frozen stock of *F. novicida*.
5. Tryptic Soy Broth (TSB).
6. Spectrophotometer.
7. Bacterial incubator at 37 °C.
8. PBS.
9. 10 % sterile cysteine in PBS.
10. Centrifuge with swing-out buckets that holds 96-well plates.
11. Gentamicin stock solution at 100 mg/mL in PBS. Store at −20 °C.
12. Cell culture facility and equipment including 37 °C, 5 % CO_2, water-jacketed cell culture incubator and laminar flow hood.
13. Standard LDH release assay.

14. Lysis solution: 10 % Triton X-100.
15. ELISA Kits for IL-1α, IL-1β or IL-18.
16. 96-Well ELISA plates.
17. Plate sealers or Parafilm.
18. 96-Well plate reader (absorbance) and analysis software.

2.5 Analysis of Caspase-1 Processing by Western Blotting

1. Tissue-culture treated 6-well plates.
2. Macrophage medium.
3. Frozen stock of *F. novicida*.
4. Tryptic Soy Broth (TSB).
5. Spectrophotometer.
6. Bacterial incubator at 37 °C.
7. PBS.
8. 10 % sterile cysteine in PBS.
9. Centrifuge with swing-out buckets that holds 6-well plates.
10. Gentamicin stock solution at 100 mg/mL in PBS. Store at –20 °C.
11. Cell culture facility and equipment including 37 °C, 5 % CO_2, water-jacketed cell culture incubator and laminar flow hood.
12. Safe-Lock 2 mL tubes (Eppendorf).
13. 100 % Trichloroacetic acid.
14. Ice.
15. Tabletop centrifuge at 4 °C.
16. Heat block at 95 °C.
17. RIPA buffer: 50 mM Tris–HCl, 150 mM NaCl, 0.1 % SDS, 0.5 % Sodium-Deoxycholate and 1 % NP-40.
18. Ice-cold acetone.
19. SDS-PAGE sample buffer.
20. 1 M Tris–HCl, pH 12.
21. SDS-PAGE and Western blot equipment
22. 14 % Tris/Glycine Gels (1.5 mm thick gels, large wells) and Tris/Glycine running buffer.
23. 0.2 μm PVDF membranes.
24. Tris-buffered Saline with 0.1 % Tween-20 (TBS-T).
25. Blocking Buffer: TBS-T with 2.5 % Skim Milk powder, 2.5 % BSA.
26. Primary and secondary-HRP coupled antibodies.
27. ECL Western blotting reagent.
28. Film, Western blot cassette, dark room, and developer.
29. Membrane stripping solution.

2.6 Visualization of Endogenous ASC Oligomers

1. Tissue-culture treated 24-well plates.
2. Sterile glass coverslips (12 mm diameter).
3. Macrophage medium.
4. Frozen stock of *F. novicida*.
5. Tryptic Soy Broth (TSB).
6. Spectrophotometer.
7. Bacterial incubator at 37 °C.
8. PBS.
9. 10 % sterile cysteine in PBS.
10. Centrifuge with swing-out buckets that holds 96-well plates.
11. Gentamicin stock solution at 100 mg/mL in PBS. Store at −20 °C.
12. Cell culture facility and equipment including 37 °C, 5 % CO_2, water-jacketed cell culture incubator and laminar flow hood.
13. Fixative: 4 % Paraformaldehyde in PBS.
14. Blocking solution: PBS with 3 % BSA, 0.1 % Saponin.
15. Primary antibodies for ASC (rat anti-ASC, clone 8E4.1, Genentech Inc. or rabbit anti-ASC, AL177, Enzo Life Sciences), bacteria, and other inflammasome components (e.g., pro-caspase-1, sc-514, Santa Cruz Biotechnology Inc.).
16. Secondary fluorophore coupled antibodies.
17. Mounting medium with DAPI.
18. Microscopy slides.
19. Hydration chamber.
20. Parafilm.

3 Methods

Ensure adequately trained persons carry out all animal procedures and that all experiments have received prior ethics approval according to your institutions requirements.

3.1 Differentiation and Long-Term Storage of Murine BMDMs

1. Fill two beakers with 100–200 mL 70 % Ethanol and move to laminar flow hood.
2. Put two forceps and one scissor in each beaker. One beaker contains "outer" tools (for touching the exterior of the mouse); the other contains "inner" tools (only for touching the mouse once fur has been removed).
3. Fill syringes with 10 mL flushing medium and attach the 26Gx1/2 needles. Usually one syringe with 10 mL medium is sufficient to flush all four bones of one mouse.

4. Euthanize a mouse by CO_2 inhalation and/or cervical dislocation.
5. Lay the mouse down on its back on a paper towel and spray the mouse with 70 % Ethanol.
6. Transfer the mouse on a dry paper towel and transfer it to the laminar flow hood.
7. *Start with outer tools*: Using the forceps, hold one leg up by its skin at the ankle. Using the scissors make a small incision into the skin and starting at this incision cut the skin open from the ankle to the belly, taking care not to cut into the peritoneum.
8. Hold the foot up with the forceps and extend the small starting incision fully around the ankle. Next, pull the skin upward over the foot and downward over the leg until the whole leg and the lower part of the body is exposed.
9. Using scissors cut the hip bone and remove the whole leg. Be careful not to cut the femur at this step. Remove the foot and place the whole leg into a sterile 10 cm petri dish.
10. Repeat the procedure on the other leg.
11. *Switch to inner tools*: Using scissors and forceps remove the muscle tissue until the bones and the knee joint are exposed.
12. Cut the leg at the knee joint to separate femur and tibia. Put the bones into a clean 10 cm petri dish.
13. To flush out the bone marrow, hold one bone in the middle with forceps and cut open both sides of the bone. The bone marrow cavity with the bright red bone marrow should now be accessible.
14. Carefully insert a needle into the bone-marrow cavity and flush out the marrow into a 50 mL tube with 2–3 mL of flushing medium per bone. If necessary turn the bone around and flush out any remaining bone marrow from the other side.
15. Repeat this process with the remaining three bones of the same mouse.
16. Centrifuge the bone marrow for 5 min at $300 \times g$, remove the supernatant and resuspend in 1 mL macrophage medium using a 1 mL pipet.
17. Add an additional 4 mL of macrophage medium and mix by gently pipetting.
18. Count the cells using a hemocytometer and determine the total number of progenitor cells. Plate 5×10^6 progenitor cells into a non-tissue culture treated 10 cm petri dish in 10 mL macrophage medium and incubate at 37 °C for 3 days.
19. After 3 days check the cells under a microscope. Part of the progenitor cells should already have settled down and are starting to form colonies, while others will still be free floating.

20. Add 4 mL of fresh pre-warmed macrophage medium directly into each petri dish without removing the medium already in the dishes, and incubate at 37 °C for 3 days (*see* **Note 1**).
21. Remove the medium containing non-adherent cells and add 5 mL of ice-cold PBS per petri dish. Leave the petri dish at 4 °C for 15–30 min.
22. Gently scrape the BMDMs with a cell scraper and transfer to a 50 mL tube. More PBS can be added to make the collection easier.
23. Centrifuge the cells for 5 min at 4 °C at $300 \times g$, remove the supernatant and resuspend the cell pellet in ice-cold FCS. Count the macrophages to determine the concentration of cells/mL.
24. Prepare a dilution of BMDMs at 2×10^8/mL in ice-cold FCS. Gently mix this dilution with ice-cold FCS containing 20 % tissue culture grade DMSO to make a suspension at 1×10^8 BMDMs/mL in FCS with 10 % DMSO.
25. Aliquot 1 mL of the final suspension into cryo vials and freeze overnight in an −80 °C freezer using an isopropanol freezing chamber.
26. Transfer to liquid nitrogen for long-term storage.

3.2 Thawing and Propagating BMDMs

1. Get the required amount of vials of frozen macrophages from liquid nitrogen long-term storage and keep them on dry ice. One vial per BMDM genotype is usually sufficient and the procedure for one vial is outlined below.
2. Quickly thaw the BMDMs in a 37 °C water bath just until the moment the cells are thawed. Spray the vials with 70 % Ethanol, dry and move to a laminar flow hood.
3. Transfer the BMDMs into a 50 mL tube.
4. Slowly add 9 mL of cold macrophage medium drop-wise, while swirling.
5. Pellet the BMDMs by centrifugation at $300 \times g$ for 5 min at 4 °C, remove the supernatant and resuspend them in 10 mL macrophage medium. This step removes most of the DMSO.
6. Transfer the BMDMs into 1 petri dish and incubate them overnight at 37 °C.
7. Check the macrophages under a microscope. Most of the BMDMs should be adherent at this time.
8. Remove the medium containing any non-adherent cells and add 5 mL of ice-cold PBS per petri dish. Leave the petri dish at 4 °C for 15–30 min.
9. Gently scrape the BMDMs with a cell scraper and transfer to a 50 mL tube. More PBS can be added to make the collection easier.

10. Centrifuge the cells for 5 min at 4 °C at 300×*g*, remove the supernatant and resuspend the cell pellet in 50 mL macrophage medium.
11. Transfer the BMDMs into 5 petri dishes and incubate at 37 °C for 2 days. After that the BMDMs are ready to be seeded for an infection as outlined below. The BMDMs can be kept up to 10 days after being thawed.

3.3 Harvesting and Pre-stimulating of BMDMs for an Infection

1. Move the petri dishes containing BMDMs to a laminar flow hood. You can expect $3–4 \times 10^8$ BMDMs per plate at this time, i.e., a total of approximately 2×10^9 BMDMs.
2. Remove the medium containing any non-adherent cells and add 5 mL of ice-cold PBS per petri dish. Leave the petri dish at 4 °C for 15–30 min.
3. Gently scrape the BMDMs with a cell scraper and transfer to 50 mL tube. More PBS can be added to make the collection easier.
4. Centrifuge the cells for 5 min at 4 °C at 300×*g*, remove the supernatant and resuspend the cell pellet in 30 mL macrophage medium.
5. Count the BMDMs twice using a hemocytometer to determine the concentration of cells/mL.
6. Prepare the desired amount of a cell suspension at 2.5×10^5 BMDMs/mL. Replate remaining BMDMs at $2–3 \times 10^8$ BMDMs/dish into petri dishes for later use. The BMDMs can be kept up to 10 days after being thawed.
7. To pre-stimulate the macrophages and induce the expression of the pro-forms of IL-1α and IL-1β, add LPS to a final concentration of 0.1 μg/mL to the cell suspensions. However, depending on the type of experiment prestimulation might not be necessary or desirable (*see* **Note 2**).
8. Depending on whether you want to analyze pyroptosis and the release of mature cytokines or the processing of caspase-1 into its p10 and p20 subunits, seed the cells according to the protocols outlined in Subheadings 3.4 and 3.5 below.

3.4 Measurement of Pyroptosis and Cytokine Maturation

1. Seed BMDMs into 96-well plates at 5×10^4 cells/well in 0.2 mL macrophage medium and incubate overnight at 37 °C. For a typical experimental setup (*see* **Note 3** and Fig. 1).
2. Start overnight cultures of *F. novicida* directly from frozen stock into TSB supplemented with 0.1 % cysteine. Grow overnight at 37 °C with constant shaking. Other pathogens might require different growth conditions to maximize inflammasome activation (*see* **Note 4**).
3. *F. novicida* do not require subculturing prior to infection. However, if your pathogen of choice requires growth to

logarithmic phase to induce expression of virulence factors (*see* **Note 5**), subculture your bacterial overnight culture 1:50–1:100 for 3–4 h before proceeding with the protocol.

4. Measure the OD600 of your bacterial culture in a spectrophotometer. Using an appropriate conversion factor determine the CFU/mL of your bacterial culture. For *F. novicida* use the following conversion: OD600 $1.0 = 3 \times 10^9$ CFU/mL (*see* **Note 6**).
5. Prepare a dilution of your bacteria in macrophage medium at the desired Multiplicity of Infection (MOI), i.e., the final number of bacteria per cell in your infection assay.
6. Remove the medium from the 96-well plate containing your BMDMs and add 0.2 mL/well of the bacterial suspension or plain macrophage medium as required by the experimental setup.
7. Centrifuge the plate for 15 min at $700 \times g$ to synchronize the infection and then incubate for 30 min at 37 °C to allow the bacteria to invade the cells.
8. Add 1/10 of well volume (20 μL) of macrophage medium containing 1 mg/mL gentamicin to reach a final concentration of 100 μg/mL in order to kill all extracellular bacteria, and incubate for 90 min at 37 °C (*see* **Note 7**).
9. Wash the infected BMDMs twice with pre-warmed macrophage medium. Add fresh pre-warmed macrophage medium containing 10 μg/mL of gentamicin.
10. Move the plates back into the incubator for the optimal incubation time. With *F. novicida* at an MOI of 100, inflammasome activation can be detected as early as 4 h post-infection (*see* **Note 6**).
11. Remove the plates from the incubator 10 min prior to the desired time-point and remove 1/10 of volume from the uninfected wells that serve as the 100 % lysis control. Add back 1/10 of volume of lysis solution and move back to the incubator for 10 min.
12. Mix the culture medium in the wells of the 100 % lysis control with a multichannel pipette.
13. For pyroptosis measurements (LDH release assay) transfer 50 μL of supernatant from each well to a new 96-well plate. Measure LDH release immediately on fresh supernatant samples according to the manufacturer's protocol.
14. Transfer the remaining supernatant from the infection plate to a new 96-well plate excluding the 100 % lysis control samples. The samples on this plate can be used immediately for IL-1α, IL-1β, or IL-18 ELISAs according to the manufacturer's instruction or frozen at −20 °C for later use.

3.5 Analysis of Caspase-1 Processing by Western Blotting

1. Seed BMDMs into 6-well plates at 1.25×10^6 cells per well in 5 mL of macrophage medium and incubate overnight at 37 °C (*see* **Note 8**).
2. Follow **steps 2–8** of Subheading 3.4 to infect the macrophages.
3. Remove the medium containing 100 μg/mL gentamicin from the macrophages. Wash the infected BMDMs twice with pre-warmed plain DMEM to remove all traces of FCS. Add 1.8 mL fresh pre-warmed, unsupplemented DMEM containing 10 μg/mL of gentamicin (*see* **Note 8**).
4. Move the plates back into the incubator for the remainder of the infection.
5. At the desired time-point, remove the plate from the incubator and transfer the supernatant (1.8 mL) to a 2 mL Safe-Lock tube. Add 0.2 mL of 100 % Trichloroacetic acid to reach a final concentration of 10 %.
6. Vortex and precipitate the proteins for at least 1 h on ice. Lysate samples can be prepared simultaneously.
7. Centrifuge the tubes for 30 min at 4 °C at full speed in a table-top centrifuge.
8. Remove the supernatant and wash the pellet with 1 mL of ice-cold acetone. Centrifuge for 10 min at 4 °C at full speed in a tabletop centrifuge.
9. Remove the supernatant and air-dry the pellet for 10–20 min at room temperature (RT).
10. Resuspend the pellet in 20–40 μL of 1× SDS-PAGE sample buffer. If the sample buffer turns yellow, re-adjust the pH by adding 1 μL of 1 M Tris–HCl pH 12.
11. Incubate the samples for 5–10 min at 95 °C. Let cool and spin down.
12. Following protein precipitation of supernatants, prepare cell lysates in RIPA buffer as follows: Transfer plates on ice after removing the supernatants for analysis.
13. Wash 2× with PBS.
14. Add RIPA buffer containing protease inhibitors according to plate format (for volumes, *see* **Note 9**) and incubate for 10 min on ice.
15. Scrape the cells and transfer to 1.5 mL tubes.
16. Incubate on ice for 30 min, with occasional vortexing.
17. Centrifuge for 10 min at 4 °C at full speed in a tabletop centrifuge.
18. Transfer the supernatant to a new tube without aspirating the pellet and add 2× SDS-PAGE sample buffer.

19. Heat the samples for 5–10 min at 95 °C, spin down and use for SDS-PAGE. The pro-forms of caspase-1, caspase-11 and IL-1β can be readily detected in cell lysates.
20. Proceed to Western blotting of precipitated supernatants and cell lysates as follows: Load the total sample of one well of a 6-well plate per lane of a 14 % SDS-PAGE and subject to electrophoresis.
21. Transfer the gel on a PVDF membrane (45 min at 100 V for BioRad MiniProtean gels).
22. Block the membrane for 1 h in blocking buffer at RT with constant shaking.
23. Add your primary antibody at an appropriate concentration in blocking buffer (*see* **Note 8**). Incubate overnight at 4 °C with constant shaking.
24. Remove the primary antibody and store at −20 °C for repeated use. Wash membranes with three quick washes and three additional 10 min washes with TBS-T.
25. Add the secondary HRP-coupled antibody at an appropriate dilution in blocking buffer and incubate for 45 min at RT, with constant shaking.
26. Remove secondary antibody. Wash membranes with three quick washes and at least four additional 10 min washes with TBS-T.
27. Tap-dry the membrane and transfer on saran wrap or parafilm. Add 1 mL ECL Western blot reagent, spread on the membrane and incubate for 5 min at RT.
28. Tap dry membrane, transfer to a Western blotting cassette and develop the blot using standard techniques and equipment.
29. To re-probe your blot with additional antibodies (*see* **Note 8**), strip the membrane with a mild stripping solution according to the manufacturer's instructions.
30. Repeat **steps 22–28** of Subheading 3.5 with different antibodies.

3.6 Visualization of Endogenous ASC Oligomers

1. Seed BMDMs into 24-well plates containing coverslips at 2.5×10^5 cells per well in 1 mL of macrophage medium and incubate overnight at 37 °C.
2. Follow **steps 2–10** of Subheading 3.4 to infect the macrophages.
3. Remove plate from incubator and wash each well 3× with PBS.
4. Remove the final wash and add 1 mL of fixative to each well and fix the cells for 15 min at 37 °C.
5. Wash each well 3× with PBS.
6. Prepare a hydration chamber, cover bottom with parafilm.

7. Transfer coverslips to hydration chamber, cover with PBS to prevent drying.
8. Remove PBS and add primary antibodies in blocking buffer (0.1 mL per coverslip) at the appropriate concentration (*see* **Note 10**). Incubate for 30 min.
9. Remove primary antibodies and wash coverslips 3× with PBS.
10. Add secondary fluorophore-coupled antibodies in blocking buffer. To outline the actin cytoskeleton, fluorophore-coupled Phalloidin (such as Alexa Fluor® 594 Phalloidin, Invitrogen) can be added to the secondary antibodies. Incubate for 30 min.
11. Remove secondary antibodies and wash coverslips 3× with PBS.
12. Mount coverslips on Microscopy slides using mounting medium with DAPI and visualize ASC specks by fluorescence microcopy or confocal microscopy.

4 Notes

1. This step ensures that the cells have enough MCSF to fully differentiate. After 6 days most of the bone marrow-derived macrophages (BMDMs) should have differentiated and are adherent to the plate.
2. Analysis of inflammasome-mediated cytokine maturation and release requires pretreatment with an appropriate TLR-ligand such as LPS, PAM3CSK or others. Pre-stimulation provides the Signal 1 of the inflammasome (*see* Subheading 1) and induces the expression of pro-IL-1α or pro-IL-1β, as well as certain inflammasome receptors (NLRP3, AIM2). In contrast, NLRC4, Nlrp1b, ASC and pro-caspase-1 are constitutively expressed. Pre-stimulation might not be necessary or desirable, since the bacterial infection itself can provide the Signal 1. For example, *F. novicida* activates TLR2 during infection, and *S. typhimurium* LPS is an excellent activator of TLR4 [32]. Thus, infection of unstimulated BMDMs with *F. novicida* will lead to the induction of AIM2 and pro-IL-1, and initiate inflammasome activation. Although inflammasome activation will be significantly delayed in unstimulated macrophages compared to pre-stimulated macrophages, this setup might be nevertheless desirable, if the goal is to study how different innate immune signaling pathways interact over the course of a pathogen infection, for example how TLR4/TRIF-signaling induces pro-caspase-11 expression [32, 33]. Unlike IL-1β maturation, pyroptosis does not require any Signal 1, if initiated by a constitutively expressed NLR, such as NLRC4 or Nlrp1b. Therefore, we recommend to pre-stimulate if analyzing inflammasome

activation by these receptors, since pyroptosis can lyse the infected cell before detectable levels of pro-IL-1 are expressed.

3. Both, pyroptosis and the release of mature cytokines can be easily detected with commercially available kits. In general, 50 μL of supernatant are required to measure the release of LDH and 100 μL are necessary for cytokine ELISA. Thus, both assays can be done from the supernatant collected from 96-well plates. IL-1β ELISAs are most commonly used, but ELISAs for IL-1α and IL-18 are available as well. The measurement of pyroptosis is based on the detection of the cytosolic Lactate Dehydrogenase (LDH), which is released during necrosis and pyroptosis, or by cell lysis with a detergent. The percentage of cell death is calculated as follows: (LDH infected − LDH uninfected/LDH total lysis − LDH uninfected) × 100. Therefore, the experimental setup requires additional uninfected wells that serve as uninfected and total lysis controls (Fig. 2). Total lysis is achieved by the addition of lysis solution (10 % Triton X-100). Use the following outline as an example on how to set up your 96-well plate experiment: For inflammasome studies, we recommend to use BMDMs from wild-type C57/BL6 mice and C57/BL6 *Casp-1-deficient* mice (aka *Casp-1/Casp-11* dKO) [31]. If your pathogen activates inflammasomes you expect to see a significant difference in LDH release or cytokine maturation between these two genotypes. Additional BMDMs of different genotypes can be added, such as *AIM2*$^{-/-}$ BMDMs for *F. novicida* infections, since *F. novicida* activates the AIM2-inflammasome in murine BMDMs [17–19]. We recommend testing different MOIs of your pathogen and unstimulated versus pre-stimulated cells to find the optimal infection conditions. All sample conditions should be done in triplicate or quadruplicate, as outlined in Fig. 2.

4. The protocol outline above describes the infection with *F. novicida*, an intracellular pathogen that activates AIM2 in murine BMDMs, a cytoplasmic DNA sensor [17–19]. However the basic infection protocol is similar for intracellular as well as extracellular pathogens.

5. The first point to consider, when planning your infection, is to find the optimal growth or induction conditions that your pathogen requires to express its virulence factors and/or successfully invade cells. The particular growth phase your bacterial culture is in can significantly change the outcome of an infection. For example, logarithmic phase *S. typhimurium* will express its SPI-1 Type 3 secretion system and inject flagellin into BMDMs, thus activating the NLRC4 inflammasome within 1–2 h [9, 29]. In contrast, stationary phase *S. typhimurium* downregulates SPI-1 expression and activates NLRP3, NLRC4 and the noncanonical caspase-11 inflammasome through other

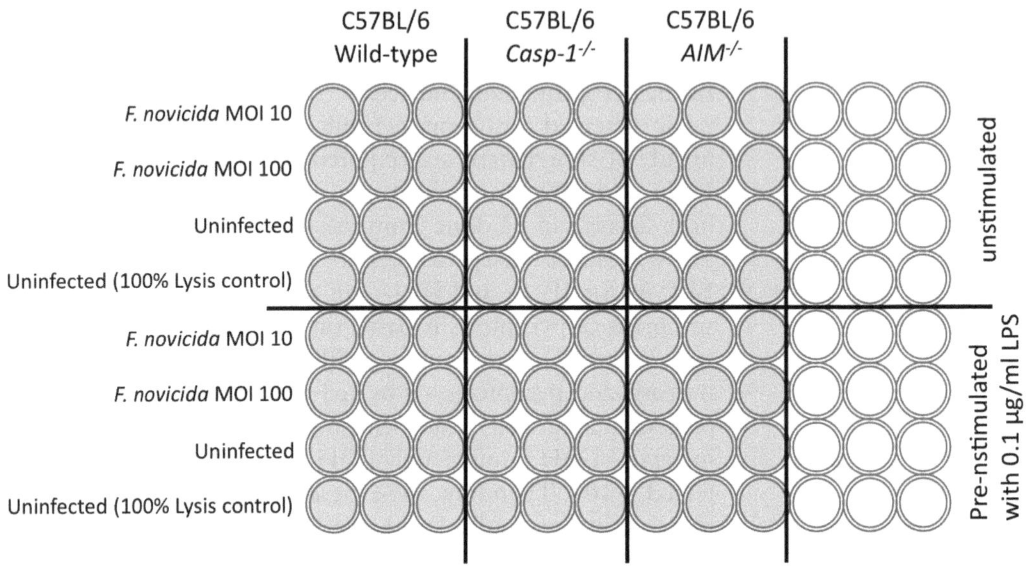

Fig. 2 Overview of an infection setup in 96-well plate format. BMDMs of the indicated genotypes are seeded at 5×10^4/well, and are either left unstimulated or pre-stimulated with 0.1 µg/mL LPS. The next day, BMDMs are infected with *F. novicida* at the indicated Multiplicity of Infection (MOI) or left uninfected. For LDH release assays, lysis solution is added to uninfected wells labeled 100 % lysis control before collecting the supernatants for analysis

virulence factors [26, 32]. Thus, understanding the particular pathogen and its virulence factors is of crucial importance to successfully study inflammasome activation.

6. Another point to consider is the MOI, i.e., the average number of bacteria per cell and the length of the infection before supernatant samples are collected. We recommend starting with a range of different MOIs and a short time-course to determine the optimal condition to measure maximal inflammasome activation. Be aware that the optimal length of infection can vary between 30 min and 24 h or more.
7. Finally, depending on the length of the infection, it is important to determine at what time point to stop bacterial uptake by killing the extracellular bacteria by the addition of gentamicin. This step is important, since many pathogens will start replicating extracellularly in the macrophage medium and would otherwise overgrow the BMDM culture. For most bacteria incubation times in the range of 30–60 min are sufficient. Opsonizing the bacteria could be used to force this process. Importantly, some bacteria activate the inflammasome so fast that the infection can be stopped before gentamicin needs to be added. An example for this is the activation of NLRC4 by logarithmic phase *S. typhimurium*, which takes only 1–2 h [9].

Importantly, long exposure to high concentrations of the membrane impermeable antibiotic gentamicin can nevertheless result in the leakage of the antibiotic into the cells, where it can affect bacterial viability. Therefore, we recommend replacing the medium containing 100 μg/mL of gentamicin (a concentration sufficient to kills extracellular bacteria) with medium containing 10 μg/mL (merely preventing bacterial growth) for longer infections.

8. The protocol outlined in Subheading 3.5 describes how to collect and concentrate supernatant samples from cells infected with *F. novicida*. Although Western blots can also be done on samples from 96-well plates, this is difficult as the amount of processed caspase-1 p10 and p20 or caspase-11 p30 in the cell supernatants is very low. Our protocol allows for a 100-fold concentration of supernatant samples, thus making the detection of processed caspase subunits significantly easier. The key to this concentration step is to precipitate the supernatant of 1–1.25 × 10^6 BMDMs. As mentioned in the protocol, it is important to collect the secreted proteins into plain, unsupplemented DMEM, since serum albumin in the medium will overload the gel after precipitation. The same protocol with minor modifications can be used for different pathogens or other inflammasome stimuli. For pathogens/stimuli that lead to a rapid activation of the inflammasome, such as logarithmic phase *S. typhimurium*, we recommend to change the protocol as follows: Wash the BMDMs twice with plain pre-warmed DMEM before the infection. Then, infect the cells with 1.8 mL of plain, unsupplemented medium containing your bacteria. After 1–2 h, collect the 1.8 mL of supernatant, still containing extracellular bacteria, and process the supernatant as outlined in Subheading 3.5. We do not recommend leaving the BMDMs too long in medium without FBS and MCSF. For very long infections (20–24 h), such the activation of the noncanonical caspase-11 inflammasome [31–33], it might be therefore advisable to start collecting the supernatant later, for example only after 8–10 h of infection. In that case, the wash steps outlined in **step 3** of Subheading 3.5 are done in macrophage medium with supplement. Then, at 8–10 h post infection, the cells are washed twice with plain pre-warmed DMEM and 1.8 mL of plain, unsupplemented medium is added to collect secreted proteins. Besides blotting for caspase-1 p10 and p20 subunits or caspase-11 p30 (clone 17D9), we routinely reprobe the membranes for IL-1α (Biovision), IL-1β (R&D Systems), and IL-18 (Abcam) in addition or as an alternative to the ELISAs described in Subheading 3.4. Additionally, the release of cytoplasmic danger signals such as HMGB1 can also be detected by Western blotting.

9. Lysate samples from 96-well and 6-well plates can be prepared in RIPA buffer using standard protocols. For addition of RIPA buffer containing protease inhibitors, use 20 μL/well in 96-well plates, 300 μL/well in 6-well plates. The pro-forms of caspase-1, caspase-11, and IL-1β can be readily detected in cell lysates.
10. The optimal concentration of primary antibodies needs to be determined empirically. We use rat anti-ASC (clone 8E4.1, Genentech Inc.) at 1:1,000, rabbit anti-ASC (AL177; Enzo Life Sciences) at 1:1,000, and rabbit anti-caspase-1 (sc-514, Santa Cruz Biotechnology Inc.) at 1:100.

Acknowledgment

We would like to thank Maikke B. Ohlson and Jens Kortmann for critical reading of the manuscript. This work was supported by awards AI095396 and AI08972 from the National Institute of Allergy and Infectious Diseases to D.M., a Stanford Digestive Disease Center (DDC) pilot grant to P.B. and a long-term fellowship from the Human Frontiers in Science Program (HFSP) to P.B.

References

1. Schroder K, Muruve DA, Tschopp J (2009) Innate immunity: cytoplasmic DNA sensing by the AIM2 inflammasome. Curr Biol 19(6):R262–R265. doi:S0960-9822(09)00677-0 [pii] 10.1016/j.cub.2009.02.011
2. Martinon F, Burns K, Tschopp J (2002) The inflammasome: a molecular platform triggering activation of inflammatory caspases and processing of proIL-beta. Mol Cell 10(2):417–426. doi:S1097276502005993 [pii]
3. Lamkanfi M (2011) Emerging inflammasome effector mechanisms. Nat Rev Immunol 11(3):213–220. doi:nri2936 [pii] 10.1038/nri2936
4. Keller M, Ruegg A, Werner S, Beer HD (2008) Active caspase-1 is a regulator of unconventional protein secretion. Cell 132(5):818–831. doi:S0092-8674(08)00111-6 [pii] 10.1016/j.cell.2007.12.040
5. Fink SL, Cookson BT (2005) Apoptosis, pyroptosis, and necrosis: mechanistic description of dead and dying eukaryotic cells. Infect Immun 73(4):1907–1916. doi:73/4/1907 [pii] 10.1128/IAI.73.4.1907-1916.2005
6. Boyden ED, Dietrich WF (2006) Nalp1b controls mouse macrophage susceptibility to anthrax lethal toxin. Nat Genet 38(2):240–244. doi:ng1724 [pii] 10.1038/ng1724
7. Hornung V, Latz E (2010) Critical functions of priming and lysosomal damage for NLRP3 activation. Eur J Immunol 40(3):620–623. doi:10.1002/eji.200940185
8. Skeldon A, Saleh M (2011) The inflammasomes: molecular effectors of host resistance against bacterial, viral, parasitic, and fungal infections. Front Microbiol 2:15. doi:10.3389/fmicb.2011.00015
9. Mariathasan S, Newton K, Monack DM, Vucic D, French DM, Lee WP, Roose-Girma M, Erickson S, Dixit VM (2004) Differential activation of the inflammasome by caspase-1 adaptors ASC and Ipaf. Nature 430(6996):213–218. doi:10.1038/nature02664 nature02664 [pii]
10. Amer A, Franchi L, Kanneganti TD, Body-Malapel M, Ozoren N, Brady G, Meshinchi S, Jagirdar R, Gewirtz A, Akira S, Nunez G (2006) Regulation of Legionella phagosome maturation and infection through flagellin and host Ipaf. J Biol Chem 281(46):35217–35223. doi:M604933200 [pii] 10.1074/jbc.M604933200
11. Franchi L, Amer A, Body-Malapel M, Kanneganti TD, Ozoren N, Jagirdar R, Inohara N, Vandenabeele P, Bertin J, Coyle A, Grant EP, Nunez G (2006) Cytosolic flagellin requires Ipaf for activation of caspase-1 and

interleukin 1beta in salmonella-infected macrophages. Nat Immunol 7(6):576–582. doi:ni1346 [pii] 10.1038/ni1346
12. Miao EA, Alpuche-Aranda CM, Dors M, Clark AE, Bader MW, Miller SI, Aderem A (2006) Cytoplasmic flagellin activates caspase-1 and secretion of interleukin 1beta via Ipaf. Nat Immunol 7(6):569–575. doi:ni1344 [pii] 10.1038/ni1344
13. Ren T, Zamboni DS, Roy CR, Dietrich WF, Vance RE (2006) Flagellin-deficient Legionella mutants evade caspase-1- and Naip5-mediated macrophage immunity. PLoS Pathog 2(3):e18. doi:10.1371/journal.ppat.0020018
14. Miao EA, Mao DP, Yudkovsky N, Bonneau R, Lorang CG, Warren SE, Leaf IA, Aderem A (2010) Innate immune detection of the type III secretion apparatus through the NLRC4 inflammasome. Proc Natl Acad Sci USA 107(7):3076–3080. doi:0913087107 [pii] 10.1073/pnas.0913087107
15. Kofoed EM, Vance RE (2011) Innate immune recognition of bacterial ligands by NAIPs determines inflammasome specificity. Nature 477(7366):592–595. doi:10.1038/nature10394 nature10394 [pii]
16. Zhao Y, Yang J, Shi J, Gong YN, Lu Q, Xu H, Liu L, Shao F (2011) The NLRC4 inflammasome receptors for bacterial flagellin and type III secretion apparatus. Nature 477(7366):596–600. doi:10.1038/nature10510 nature10510 [pii]
17. Fernandes-Alnemri T, Yu JW, Juliana C, Solorzano L, Kang S, Wu J, Datta P, McCormick M, Huang L, McDermott E, Eisenlohr L, Landel CP, Alnemri ES (2010) The AIM2 inflammasome is critical for innate immunity to Francisella tularensis. Nat Immunol 11(5):385–393. doi:ni.1859 [pii] 10.1038/ni.1859
18. Rathinam VA, Jiang Z, Waggoner SN, Sharma S, Cole LE, Waggoner L, Vanaja SK, Monks BG, Ganesan S, Latz E, Hornung V, Vogel SN, Szomolanyi-Tsuda E, Fitzgerald KA (2010) The AIM2 inflammasome is essential for host defense against cytosolic bacteria and DNA viruses. Nat Immunol 11(5):395–402. doi:ni.1864 [pii] 10.1038/ni.1864
19. Jones JW, Kayagaki N, Broz P, Henry T, Newton K, O'Rourke K, Chan S, Dong J, Qu Y, Roose-Girma M, Dixit VM, Monack DM (2010) Absent in melanoma 2 is required for innate immune recognition of Francisella tularensis. Proc Natl Acad Sci USA 107(21):9771–9776. doi:1003738107 [pii] 10.1073/pnas.1003738107
20. Kim S, Bauernfeind F, Ablasser A, Hartmann G, Fitzgerald KA, Latz E, Hornung V (2010) Listeria monocytogenes is sensed by the NLRP3 and AIM2 inflammasome. Eur J Immunol 40(6):1545–1551. doi:10.1002/eji.201040425
21. Warren SE, Armstrong A, Hamilton MK, Mao DP, Leaf IA, Miao EA, Aderem A (2010) Cutting edge: cytosolic bacterial DNA activates the inflammasome via Aim2. J Immunol 185(2):818–821. doi:jimmunol.1000724 [pii] 10.4049/jimmunol.1000724
22. Wu J, Fernandes-Alnemri T, Alnemri ES (2010) Involvement of the AIM2, NLRC4, and NLRP3 inflammasomes in caspase-1 activation by Listeria monocytogenes. J Clin Immunol 30(5):693–702. doi:10.1007/s10875-010-9425-2
23. Tsuchiya K, Hara H, Kawamura I, Nomura T, Yamamoto T, Daim S, Dewamitta SR, Shen Y, Fang R, Mitsuyama M (2010) Involvement of absent in melanoma 2 in inflammasome activation in macrophages infected with Listeria monocytogenes. J Immunol 185(2):1186–1195. doi:jimmunol.1001058 [pii] 10.4049/jimmunol.1001058
24. Sauer JD, Witte CE, Zemansky J, Hanson B, Lauer P, Portnoy DA (2010) Listeria monocytogenes triggers AIM2-mediated pyroptosis upon infrequent bacteriolysis in the macrophage cytosol. Cell Host Microbe 7(5):412–419. doi:S1931-3128(10)00110-1 [pii] 10.1016/j.chom.2010.04.004
25. Masumoto J, Taniguchi S, Ayukawa K, Sarvotham H, Kishino T, Niikawa N, Hidaka E, Katsuyama T, Higuchi T, Sagara J (1999) ASC, a novel 22-kDa protein, aggregates during apoptosis of human promyelocytic leukemia HL-60 cells. J Biol Chem 274(48):33835–33838
26. Broz P, Newton K, Lamkanfi M, Mariathasan S, Dixit VM, Monack DM (2010) Redundant roles for inflammasome receptors NLRP3 and NLRC4 in host defense against Salmonella. J Exp Med 207(8):1745–1755. doi:jem.20100257 [pii] 10.1084/jem.20100257
27. Fernandes-Alnemri T, Wu J, Yu JW, Datta P, Miller B, Jankowski W, Rosenberg S, Zhang J, Alnemri ES (2007) The pyroptosome: a supramolecular assembly of ASC dimers mediating inflammatory cell death via caspase-1 activation. Cell Death Differ 14(9):1590–1604. doi:4402194 [pii] 10.1038/sj.cdd.4402194
28. Bryan NB, Dorfleutner A, Rojanasakul Y, Stehlik C (2009) Activation of inflammasomes requires intracellular redistribution of the apoptotic speck-like protein containing a caspase recruitment domain. J Immunol 182(5):3173–3182. doi:182/5/3173 [pii] 10.4049/jimmunol.0802367
29. Broz P, von Moltke J, Jones JW, Vance RE, Monack DM (2010) Differential requirement

for Caspase-1 autoproteolysis in pathogen-induced cell death and cytokine processing. Cell Host Microbe 8(6):471–483. doi:S1931-3128(10)00380-X [pii] 10.1016/j.chom.2010.11.007

30. Miao EA, Leaf IA, Treuting PM, Mao DP, Dors M, Sarkar A, Warren SE, Wewers MD, Aderem A (2010) Caspase-1-induced pyroptosis is an innate immune effector mechanism against intracellular bacteria. Nat Immunol 11(12):1136–1142. doi:ni.1960 [pii] 10.1038/ni.1960

31. Kayagaki N, Warming S, Lamkanfi M, Vande Walle L, Louie S, Dong J, Newton K, Qu Y, Liu J, Heldens S, Zhang J, Lee WP, Roose-Girma M, Dixit VM (2011) Non-canonical inflammasome activation targets caspase-11. Nature 479(7371):117–121. doi:10.1038/nature10558 nature10558 [pii]

32. Broz P, Ruby T, Belhocine K, Bouley DM, Kayagaki N, Dixit VM, Monack DM (2012) Caspase-11 increases susceptibility to Salmonella infection in the absence of caspase-1. Nature 490(7419):288–291. doi:10.1038/nature11419 nature11419 [pii]

33. Rathinam VA, Vanaja SK, Waggoner L, Sokolovska A, Becker C, Stuart LM, Leong JM, Fitzgerald KA (2012) TRIF licenses caspase-11-dependent NLRP3 inflammasome activation by gram-negative bacteria. Cell 150(3):606–619. doi:S0092-8674(12)00825-2 [pii] 10.1016/j.cell.2012.07.007

Chapter 7

Detection of Pyroptosis by Measuring Released Lactate Dehydrogenase Activity

Manira Rayamajhi, Yue Zhang, and Edward A. Miao

Abstract

Pyroptosis is a form of programmed, inflammatory cell death that is dependent on the activation of a cysteine protease caspase-1. Following caspase-1 activation via inflammasomes (including NLRP3, NLRC4, Nlrp1b, and AIM2), cells lose membrane integrity and lyse, releasing lactate dehydrogenase (LDH) that is normally maintained within the cell cytosol. Thus, pyroptosis is distinct from apoptosis, which results in cellular contents being enclosed within membrane blebs during cellular demise. LDH is only released from apoptotic blebs after secondary necrosis occurs. Pyroptosis is distinct from necrosis in that it requires the activity of caspase-1. In this chapter, we describe enzymatic assays for the detection of LDH released by pyroptotic cells using a commercially available kit, as well as a simple and cost-effective method adapted from Decker et al. (J Immunol Methods 115:61–69, 1988).

Key words Pyroptosis, LDH, Inflammasome, Caspase-1

1 Introduction

Cellular response to pathogens is initiated through the engagement and recognition of unique pathogen associated molecular patterns (PAMPs) via pattern recognition receptors (PRRs) such as toll like receptors (TLRs) and Nod-like receptors (NLRs). Upon activation, a subset of NLRs forms a multiprotein complex, known as the inflammasome. Inflammasomes form a platform for the cleavage of pro-caspase-1 into its active caspase-1 form. Caspase-1 activation through NLRP3, NLRC4, Nlrp1b, and AIM2 has been shown to induce pyroptosis. Furthermore, caspase-1 mediates the proteolytic activation and release of two inflammatory cytokines, IL-1β and IL-18. NLRC4 and Nlrp1b are expressed in sufficient levels in bone-marrow-derived macrophages and dendritic cells, whereas NLRP3 expression is not sufficient at basal levels, and requires priming via TLRs or TNF [1, 2]. AIM2 expression is induced by type I interferon signaling [3]. In human monocytes, cellular priming greatly enchances the response to NLRP3 activating agents.

Christine M. De Nardo and Eicke Latz (eds.), *The Inflammasome: Methods and Protocols*, Methods in Molecular Biology, vol. 1040, DOI 10.1007/978-1-62703-523-1_7, © Springer Science+Business Media New York 2013

However, NLRP3 signaling is short circuited by secretion of ATP in response to TLR agonists, and thus larger amounts of LPS (in the ng/mL range) are sufficient to induce NLRP3 inflammasome activation without the need for the addition of a second stimulus [4, 5]. Over the years, a variety of microbial and danger associated ligands capable of inducing pyroptosis via inflammasomes, have been identified. The most well studied activation mechanism is that of the NLRC4 inflammasome, which becomes active upon cytosolic delivery of flagellin, leading to caspase-1 activation and cell death via pyroptosis [6].

Pyroptotic cell death is distinct from other known forms of cell death such as apoptosis, autophagy, and necrosis in its dependence on caspase-1 [6]. While caspase-1 dependency is the hallmark of pyroptosis, a recent report suggested that caspase-11 can also trigger pyroptosis [7]. Pyroptosis has been best described in macrophages and dendritic cells. While the exact mechanisms of pyroptosis remain unclear, it is known that caspase-1 activation results in rapid formation of membrane pores, significant cell swelling followed by membrane rupture, and release of intracellular content into the extracellular space, including cytosolic proteins such as lactate dehydrogenase (LDH). Of note, after strong stimulation of inflammasomes, the entire pyroptosis process can occur in less than 1 h. Compounds such as glycine and polyethylene glycol can block membrane rupture and release of material from pyroptotic cells [8] through poorly understood mechanisms.

Here we describe a simple method to measure the release of LDH into culture supernatants during pyroptosis. Amongst the myriad of released cytosolic proteins, LDH is used as a proxy for cell death only because of the ease of measuring its enzymatic activity. LDH is a key glycolytic enzyme that oxidizes lactate to pyruvate, transferring electrons to the cofactor nicotinamide adenine dinucleotide (NAD^+) to form NADH. Both of the methods discussed here utilize the ability of NADH to donate electrons to FAD^+ complexed with the enzyme diaphorase (purified diaphorase contains FAD^+). Diaphorase is a ubiquitous enzyme that detoxifies small molecules, including quinolones. Furthermore, it is known to catalyze the reduction of tetrazolium salt (INT) into formazan. While INT is yellow, formazan is red, absorbing light at 492 nm. The increased absorption is readily detectable with a spectrophotometer (*see* Fig. 1). Thus, the amount of light absorbed is directly proportional to the number of lysed cells or cells with compromised membrane permeability. Therefore, early LDH measurement is indicative of pyroptotic cell death.

However, LDH measurement alone is not sufficient to qualify cell death as pyroptosis. The dependence of caspase-1 is absolutely essential to distinguish pyroptotic cell death from necrotic cell death or apoptosis followed by secondary necrosis (lysis of apoptotic blebs). Although the caspase-1 inhibitor (z-YVAD-fmk) and

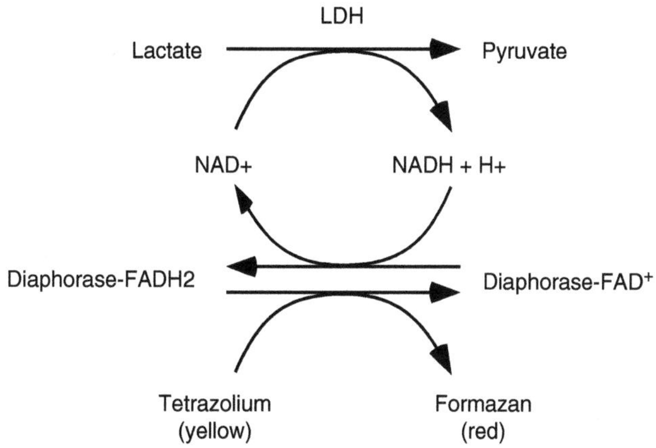

Fig. 1 Schematic of assay for LDH activity. LDH oxidizes lactate to pyruvate, reducing NAD^+ to NADH. NADH is then oxidized back to NAD^+, while reducing the diaphorase-FAD^+ complex to diaphorase $FADH_2$. Diaphorase $FADH_2$ then reduces tetrazolium to formazan

shRNA/siRNA knockdown have been extensively used in the literature, we recommend the use of cells derived from caspase-1 knockout mice wherever possible. It should be noted, however, that the commonly used caspase-1 knockout strains also lack caspase-11 [7]. Inhibitors or knock-down methods carry the potential caveats of off target effects and incomplete inhibition at later time-points. For example, reduction of caspase-1 activity by 90 % could still result in the binary event of cell death after longer incubation periods. Hence, experiments using inhibitors or knockdown should examine cell death at several time-points.

The method described below is routinely used in our laboratory to assess pyroptotic cell death in bone-marrow-derived macrophages (BMDMs) in response to *Salmonella typhimurium* infection. However, it can be adapted to assess pyroptosis in any macrophage or dendritic cell type in response to the desired known or unknown stimuli. We describe the use of a commercially available LDH assay kit as well as a noncommercial method for the determination of pyroptosis adapted from Decker et al. [9]. The latter method can be performed at approximately one third of the cost of the commercial kit.

2 Materials

1. Assay substrate mix: 12 mg/mL lactate, 0.66 mg/mL Iodonitrotetrazolium chloride (INT), 4.5 units/mL Diaphorase, 0.01 % BSA, 0.4 % sucrose in PBS. Freeze in 12 mL aliquots (*see* **Notes 1** and **2**).

2. 100× NAD^+ stock solution: 100 mg/mL in ddH_2O (working concentration is 1 mg/mL). Freeze in single use aliquots (*see* **Note 3**).
3. Lysis buffer: 9 % Triton-X100 in dH_2O. Keep at 4 °C until use.
4. Stop solution: 1 M acetic acid. Store at 4 °C or room temperature.
5. Complete medium: DMEM supplemented with 10 % FBS, 10 % L-cell conditioned medium.
6. CytoTox 96® Non-Radioactive Cytotoxicity Assay (Promega).
7. Bone-Marrow-Derived Macrophages (BMDMs) or dendritic cells (BMDCs).
8. 96-well flat bottom plates, half volume wells (Costar).
9. Plate reader.

3 Methods

3.1 Infection Conditions

1. Plate BMDMs in a 96-well plate at a density of 5×10^4 cells per well, in 100 μL of complete medium, in triplicates (*see* **Note 4**).
2. Infect cells with *Salmonella typhimurium* for the desired amount of time. The plates can be centrifuged at 500 ×*g* for 5 min at room temperature to ensure uniform infectivity and to correct for amotile strains (*see* **Note 5**).
3. Add 10 μL of lysis buffer to lyse the cells in the lysis wells. Pipette up and down to ensure that all cells have lysed.
4. Collect supernatants for cytotoxicity assay (*see* **Note 6**).

3.2 Cytotoxicity Assay

3.2.1 LDH Assay, Commercial Kit (CytoTox 96®)

1. If using a commercially available kit, follow the manufacturer's instructions. We routinely scale down to half the volumes suggested (25 μL of cell supernatant, 25 μL of reagent, and 25 μL of stop solution in half volume 96-well plates). Development takes about 8 min.

3.2.2 LDH Assay, Noncommercial

1. Thaw assay substrate mix in a room temperature water bath. Once the assay substrate is fully thawed and equilibrated to room temperature, add 25 μL of 100× NAD^+ stock to 2.5 mL of assay reagent mix. This is enough substrate mix for one 96-well plate.
2. Transfer 25 μL/well of your sample to the flat bottom assay plates.
3. Add 25 μL/well of assay substrate (containing the NAD^+) using a multichannel pipette.
4. Incubate at room temperature for about 15 min (*see* **Note 7**).
5. To stop the reaction, add 25 μL of the stop solution to each well.

6. Read the absorbance at 490 or 492 nm using appropriate plate reader (*see* **Note 8**).
7. Calculate cytotoxicity (%LDH release) using the following formula:

$$\%\text{LDH release} = \frac{(\text{Sample-Media}) - (\text{Untreated Sample} - \text{Media})}{(\text{Lysis} - \text{Lysis Media}) - (\text{Untreated Sample} - \text{Media})} \times 100$$

Sample—experimental wells, Media—cell culture media (background absorbance), Untreated sample—supernatants from uninfected or untreated cells, Lysis—supernatants from cells lysed using Triton-X, Lysis Media—cell culture media + lysis buffer

4 Notes

1. 20 mg/mL stock solutions of INT in DMSO can be stored at -20 °C.
2. We routinely use Diaphorase from Worthington Biochemical (purified from *Clostridium kluyveri*; catalog number LS004327; sold as 1 ku, or 1 kilo unit = 1,000 units; light sensitive. Diaphorase from other vendors is considerably more expensive).
3. NAD^+ is light sensitive and should be stored at −80 °C. Freeze–thaw should be strictly avoided. NAD^+ degrades at neutral pH, especially in the presence of phosphate. We have found that NAD^+ stock solutions in PBS stored at −20 °C degrade in 1 week.
4. Reserve BMDM seeded wells for untreated and lysis controls, plus empty wells for media only controls and media + lysis buffer.
5. Serum free medium or medium containing 5 % serum can be used to reduce background signal arising from LDH enzyme that is present within FBS.
6. Supernatants can be frozen at −80 °C without loss in LDH. However, ensure that all reagents and sample supernatants are at room temperature before carrying out the assay.
7. The incubation time may vary depending on the number of cells plated. When the lysis wells change color to deep maroon, the reaction can be stopped.
8. Air bubbles in the wells can interfere with the absorbance readings. Bubbles can be popped by spraying the plate lightly with 70 % ethanol or by lancing each bubble with a needle.

References

1. Franchi L, Eigenbrod T, Núñez G (2009) Cutting edge: TNF-alpha mediates sensitization to ATP and silica via the NLRP3 inflammasome in the absence of microbial stimulation. J Immunol 183:792–796
2. Bauernfeind FG et al (2009) Cutting edge: NF-{kappa}B activating pattern recognition and cytokine receptors license NLRP3 inflammasome activation by regulating NLRP3 expression. J Immunol 183:787–791
3. Jones JW et al (2010) Absent in melanoma 2 is required for innate immune recognition of Francisella tularensis. Proc Natl Acad Sci USA 107:9771–9776
4. Netea MG et al (2009) Differential requirement for the activation of the inflammasome for processing and release of IL-1beta in monocytes and macrophages. Blood 113: 2324–2335
5. Piccini A et al (2008) ATP is released by monocytes stimulated with pathogen-sensing receptor ligands and induces IL-1beta and IL-18 secretion in an autocrine way. Proc Natl Acad Sci USA 105:8067–8072
6. Miao EA, Rajan JV, Aderem A (2011) Caspase-1-induced pyroptotic cell death. Immunol Rev 243:206–214
7. Kayagaki N et al (2011) Non-canonical inflammasome activation targets caspase-11. Nature 479:117–121
8. Bergsbaken T, Fink SL, Cookson BT (2009) Pyroptosis: host cell death and inflammation. Nat Rev Microbiol 7:99–109
9. Decker T, Lohmann-Matthes ML (1988) A quick and simple method for the quantitation of lactate dehydrogenase release in measurements of cellular cytotoxicity and tumor necrosis factor (TNF) activity. J Immunol Methods 115:61–69

Chapter 8

ASC Speck Formation as a Readout for Inflammasome Activation

Andrea Stutz, Gabor L. Horvath, Brian G. Monks, and Eicke Latz

Abstract

All inflammasomes require the adapter protein apoptosis associated speck-like protein containing a CARD (ASC) for the activation of caspase-1. After inflammasome activation, ASC assembles into a large protein complex, which is termed "speck". ASC specks can be observed as they reach a size of around 1 μm and in most cells only one speck forms upon inflammasome activation. Hence, ASC speck formation can be used as a simple upstream readout for inflammasome activation. Here, we describe a method for analyzing inflammasome activation by ASC speck visualization. First, we describe the generation of a clonal inflammasome reporter macrophage cell line overexpressing fluorescently tagged ASC. We then discuss stimulation conditions and the microscopic evaluation of ASC speck formation.

Key words Inflammasome, ASC, Protein speck, NLRP3, AIM2, NLRC4, NLRP6, Microscopy, Quantitative image analysis

1 Introduction

Inflammasome activation results in the recruitment of the adaptor protein ASC, which serves as an activation platform for the inflammatory pro-caspase-1. ASC was originally found as a protein responsible for morphological changes during apoptosis. In untreated cells, ASC is localized to the cytoplasm and nucleus of cells [1]. In response to inflammasome activation, one large ASC speck (around 1 μm in diameter) assembles in a paranuclear area of the cell. Since ASC specks are large enough to be visualized by light microscopy and most cells only contain one ASC speck after inflammasome activation, their formation can thus be used as a convenient readout for inflammasome activation upstream of caspase-1 and the secretion of IL-1β. After speck formation, most of the ASC protein that had been soluble and localized throughout the cytoplasm before activation is located in the speck (*see* Fig. 1).

Christine M. De Nardo and Eicke Latz (eds.), *The Inflammasome: Methods and Protocols*, Methods in Molecular Biology, vol. 1040, DOI 10.1007/978-1-62703-523-1_8, © Springer Science+Business Media New York 2013

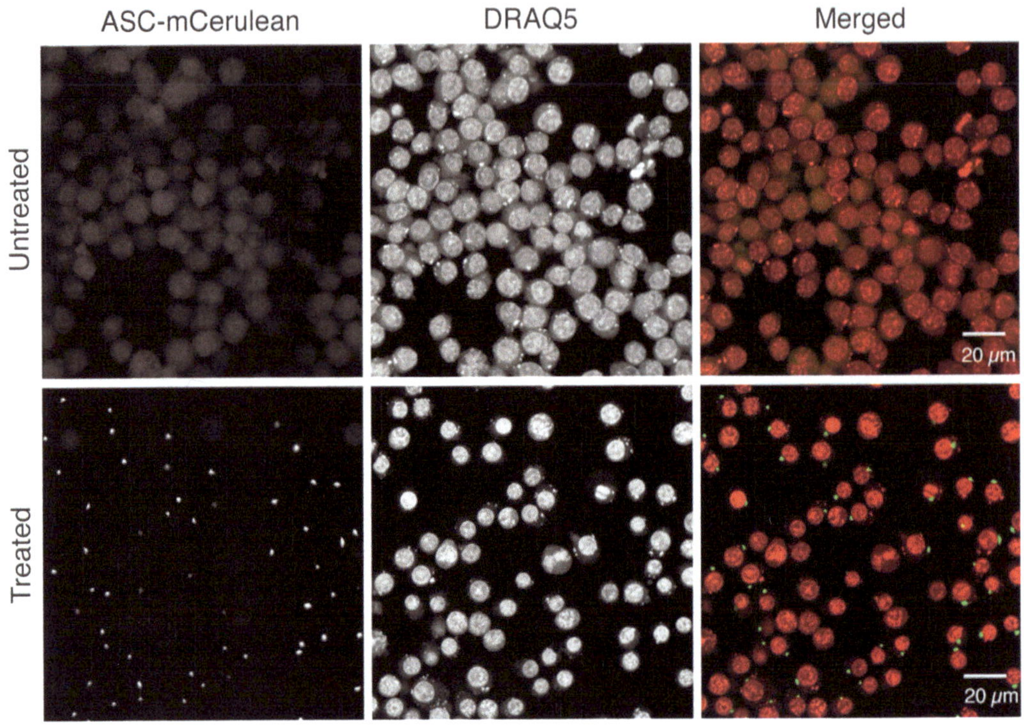

Fig. 1 Typical result of untreated and treated cells. NLRP3-FLAG overexpressing immortalized NLRP3 KO murine macrophages were transduced with ASC-mCerulean and made clonal as described in this chapter. The cells were left untreated (*top panel*) or treated with 10 μM Nigericin for 90 min (*bottom panel*), fixed and the nuclei were stained with DRAQ5. Z-stack images were taken with the Leica TCS SMD FCS confocal microscope (63× objective). Maximum projection images are shown. Note the broad, dim distribution of ASC-mCerulean in the *top panel* and the bright, small speck-like distribution of ASC in the *lower panel*

Therefore, the change of distribution can be visualized with a fluorescently tagged version of ASC. Upon inflammasome activation, the fluorescent ASC changes from a weak diffuse signal present throughout the cell (including the nucleus) to one small and localized bright spot in an activated cell. This dramatic change in fluorescence signal can be easily quantified by applying image analysis using simple thresholding algorithms. In the non-activated cell, the fluorescent signal can be set below a given threshold, and due to the large increase in fluorescence intensity in a single speck, the signal in the activated cell is regularly above the chosen threshold. In a second step, the objects lying above threshold need to be counted for a measure of activation.

In this chapter, we describe the generation of a clonal cell line overexpressing a fluorescently tagged version of ASC and describe the stimulation conditions as well as microscopic analysis. Furthermore, recommendations are given for counterstaining measures that allow normalization.

2 Materials

The following procedures require the generation of polytropic retroviral vectors. Experimentation with these vectors can pose a health risk and restrictions in use of these materials in laboratories apply. Wear appropriate personal protective equipment at all times and follow your local institution's guidelines for handling hazardous materials.

2.1 Generation of an ASC-Overexpressing Cell Line

1. 24-well and 96-well tissue culture plates.
2. Complete DMEM: DMEM (high glucose), supplemented with 10 % fetal bovine serum (FBS), optional: antibiotics, e.g., 10 μg/mL Ciprofloxacin.
3. Serum-free DMEM (SFM).
4. FBS.
5. Virus-producing cell line: HEK 293T (*see* **Note 1**).
6. Recipient cell line: immortalized WT mouse macrophages or other of your choice (*see* **Note 2**).
7. Virus packaging plasmids (e.g., pCMV-Gag-Pol, pCMV-VSV-G from Cell Biolabs) and pRP_ASC-LmCerulean retroviral plasmid (deposited at Addgene under Latz lab) (*see* **Note 3**).
8. Transfection reagent, e.g., GeneJuice (Merck).
9. 0.45 μm syringe filter with Luer lock (Polyethersulfone or similar) (*see* **Note 4**).
10. 2 mL syringe (or larger volume) with Luer Lock compatible to filter.
11. Cell culture facility with an incubator maintained at 37 °C and 5 % CO_2.

2.2 Selecting Suitable Clones: Crude Selection

1. 24-well and 6-well tissue culture plates.
2. 75 cm^2 tissue culture flasks.
3. Complete DMEM as in **item 2** of Subheading 2.1.
4. ASC-mCerulean transduced cell line as generated in Subheading 3.1.
5. Non-transduced recipient cell line from **item 6** of Subheading 2.1 as a control.
6. Option 1: FACS sorter with filter settings for CFP, e.g., 445 nm laser with 510/80 nm emission optics.
7. Option 2: Puromycin solution. Store at −20 °C in the dark. Puromycin is toxic, so handle with care.
8. Cell culture facility with an incubator maintained at 37 °C and 5 % CO_2.

2.3 Selecting Suitable Clones: Limiting Dilution

1. 96-well and 48-well tissue culture plates.
2. 75 cm^2 tissue culture flasks.
3. Complete DMEM as in **item 2** of Subheading 2.1.
4. 0.22 μm filter device.
5. Epifluorescence microscope with a filter for CFP (e.g., Filter Set 47 from Zeiss: BP 436/20, FT 455, BP 480/40).
6. Cell culture facility with an incubator maintained at 37 °C and 5 % CO_2.

2.4 Selecting Suitable Clones: Testing of Clones

1. 24-well and 6-well tissue culture plates.
2. 75 cm^2 tissue culture flasks.
3. Complete DMEM as in **item 2** of Subheading 2.1.
4. Flow cytometer with filter settings for CFP, e.g., 445 nm laser with 510/80 nm emission optics.
5. Epifluorescence microscope with a filter for CFP as in **item 5** of Subheading 2.3.
6. Different inflammasome stimuli (*see* **Note 5**).
7. Cell culture facility with an incubator maintained at 37 °C and 5 % CO_2.

2.5 Stimulation, Fixation and Counterstaining

1. Tissue culture treated plates/dishes for imaging (with flat and transparent bottom).
2. Inflammasome stimuli (*see* **Note 5**).
3. 16 % Formaldehyde in H_2O (*see* **Note 6**). Store in the dark.
4. DRAQ5. Store at 4 °C in the dark. DRAQ5 is toxic, so handle with care.

2.6 Imaging and Imaging Analysis

1. Epifluorescence microscope with a filter for CFP as in **item 5** of Subheading 2.3, and a filter for DRAQ5 (e.g., Filter Set 50 from Zeiss or similar: excitation BP 640/30, FT 660, emission BP 690/50) and 10× or 20× objective or Confocal microscope with laser lines 458 nm (or 405 nm) and 647 nm (or 633 nm) and appropriate filters as the emission band-pass filters above.
2. ImageJ [2]: available as freeware from http://rsbweb.nih.gov/ij/ (*see* **Note 7**).
3. Plugin for ImageJ: "ITCN: Image-based Tool for Counting Nuclei" available from http://www.bioimage.ucsb.edu/automatic-nuclei-counter-plug-in-for-imagej or http://imagej.nih.gov/ij/plugins/itcn.html

3 Methods

Polytropic retroviruses can transduce cells from many species, including humans. Carry out all steps of the protocol in accordance with your local institution's guidelines. Wear protective equipment and dispose of waste according to the local regulations.

3.1 Generation of Retroviral Particles for Transduction

1. On day 1, plate HEK 293T cells in a 96-well plate at 2.2×10^5 cells/mL in 100 μL/well (i.e., 2.2×10^4 cells/well) in complete DMEM (*see* **Note 8**) and incubate at 37 °C overnight. Seed at least six wells for every cell line you would like to transduce.
2. On day 2, transfect the HEK293T cells with the constructs necessary to package the retroviral vector into virus particles. In each well, 133 ng retroviral vector plasmid, 45 ng gag-pol, and 22 ng VSV-G should be transfected. Prepare the appropriate amount of DNA in 10 μL SFM per well. Also prepare 0.8 μL GeneJuice in 10 μL SFM per well. Combine the two solutions after 5 min. Wait for 20 min. Add the 20 μL transfection mix to each well and incubate at 37 °C overnight.
3. On day 3, in the morning, carefully change the medium on the virus-producing cells to 200 μL pre-warmed DMEM containing 30 % FBS (*see* **Note 9**) and incubate at 37 °C overnight.
4. In the evening, plate recipient cells in complete DMEM in a 24-well plate at around 30 % confluency (e.g., 3×10^4 immortalized WT mouse macrophages per well) and incubate at 37 °C overnight.
5. On day 4, approximately 48 h after transfection, harvest virus-containing supernatant from the 96-well plate. Use a 0.45 μm filter and a syringe to filter out the cells and debris from virus-containing supernatant.
6. Remove the medium off the recipient cells and transduce them with 1 mL of viral supernatant/well and incubate at 37 °C overnight.
7. On day 5, replace the viral supernatant on the recipient cells with 1 mL of complete DMEM and incubate the cells at 37 °C for 72 h in complete DMEM (passaging them as necessary).

3.2 Selecting Suitable Clones: Crude Selection

Two options are described: If fluorescence-activated cell sorting (FACS) is available to you, Option 1 is preferred (*see* **Note 10**). If not, proceed to Option 2 (antibiotic selection).

Option 1

1. Expand the transduced cells from **step** 7 of Subheading 3.1 to at least a 75 cm^2 flask.

2. Harvest the transduced cells and the original, non-transduced cell line and get them ready for the FACS procedure.
3. Use the original, non-transduced cell line as a control to set the gates, and select for mCerulean-positive cells (*see* **Note 11**).

Option 2

1. Prior to starting the experiment, perform a kill curve experiment to find the correct concentration of selection antibiotic to use (*see* **Note 12**): Seed $3–8 \times 10^4$ non-transduced control cells in a 24-well plate and incubate at 37 °C overnight (*see* **Note 13**). Then, replace with fresh complete DMEM containing 0–10 μg/mL puromycin (e.g., 0, 1.25, 2.5, 5, 10 μg/mL). For some cells, higher concentrations might be necessary, so you can extend your kill curve up to 25 μg/mL. Grow the cells at 37 °C for 1–4 days depending on their growth behavior. After 1–4 days, evaluate the cell survival. For the selection procedure as described below, use the lowest concentration of puromycin that killed all cells.
2. 72 h post transduction, start the antibiotic selection on the cells from **step 7** of Subheading 3.1, by adding puromycin in complete DMEM at the concentration and for the duration that you have determined in **step 1** of Option 2. Use non-transduced cells as a control for the selection.
3. On the following day, change the media to remove dead cells and/or passage the cells depending on their confluency. Proceed like this every day until the cell line has been in selection for 4 days (*see* **Note 14**).

3.3 Selecting Suitable Clones: Limiting Dilution

1. After both selection options, passage the cells twice or until they are expanded to a 75 cm^2 flask. This is also a good time point to freeze down the polyclonal cells.
2. For the limiting dilution, harvest the cells, but save the media in which the cells grew (=conditioned medium).
3. Filter the conditioned medium through a 0.22 μm filter to remove cells and debris.
4. After harvesting, wash the cells once with complete DMEM.
5. Dilute cells to 5 cells/mL in 50 mL of limiting dilution medium (use all the conditioned medium you have generated in **step 3** of Subheading 3.3, typically 15 mL, and make up to 50 mL with complete DMEM).
6. Plate 100 μL/well in 5 × 96-well plates (statistically, this results in 0.5 cells/well) (*see* **Note 15**) and incubate at 37 °C for 5–7 days.
7. After 5–7 days, screen the plates for clones using an epifluorescence microscope with filters suitable for mCerulean detection (*see* **Note 16**).

8. Pick ASC-mCerulean positive clones and transfer them to 48-well plates and proceed to Subheading 3.4. Pick as many clones as possible since many will show unwanted behavior. We generally pick around 30 clones.

3.4 Selecting Suitable Clones: Testing of Clones

1. Expand the clones from **step 8** of Subheading 3.3 to a 24-well plate. Discard clones that exhibit more than 5 % of cells with ASC specks when non-stimulated as judged by eye with a fluorescent microscope.
2. Analyze the clones by flow cytometry on mCerulean fluorescence. The clone should appear as a narrow single peak in a histogram of mCerulean. Exclude clones that have a broad distribution or even multiple populations as this might indicate that your cell line is polyclonal (*see* **Note 17**).
3. Test multiple different known stimuli (*see* **Note 18**). To this, seed 4×10^4 cells/well in 96-well plates (one well for control and another well for each stimulus).
4. To ensure that a single clone was picked, repeat the limiting dilution as in Subheading 3.3 with your successful clone with the following alterations: One 96-well plate is sufficient. Pick only 5–10 clones and test them again as described in **steps 2** and **3** of Subheading 3.4 (*see* **Note 17**).

3.5 Stimulation, Fixation and Counterstaining

1. Plate the reporter cells from **step 4** of Subheading 3.4 1 day prior to the stimulation to allow them to attach and incubate at 37 °C overnight. They should be approx. 70–80 % confluent at the time of stimulation (e.g., 1.25×10^4 cells/well in a 384-well plate or 4×10^4 cells/well in a 96-well plate).
2. Stimulate the cells using your preferred stimuli (*see* **Note 5**).
3. After stimulation, fix and counterstain the cells in one step. To the 16 % formaldehyde, add DRAQ5 at a concentration of 40 μM. Use this as a 16× stock, e.g., to 100 μL of supernatant, add 6.7 μL of the 16× stock. This will result in a final concentration of 1 % formaldehyde and 2.5 μM DRAQ5. Do not wash or change the supernatant (*see* **Notes 19** and **20**).
4. Store plates at 4 °C in the dark. The plates can be imaged up to 1 week after fixation.

3.6 Imaging and Imaging Analysis

1. Epifluorescence: Use the 10× or 20× objective with filters for CFP and DRAQ5. *See* **item 1** in Subheading 2.6 for filter details. Confocal microscope: Use the 10× or 20× objective. Use a laser with 458 nm wavelength for excitation of the mCerulean (405 nm is also possible). Use a laser with 647 nm wavelength for excitation of the DRAQ5 staining (633 nm is also possible). Take Z-stacks of the cells because cells with specks will often round up and the specks will be located further

away from the bottom of the plate. Use the maximum intensity projection of the Z-stack images for analysis.

2. During acquisition, make sure to use the same imaging parameters (gain, light intensity, pinhole, etc.) for all conditions.
3. Export images as grayscale 8-bit images (each channel separately).
4. Analysis of the ASC speck picture with ImageJ: Open at least one positive control and at least one negative control image in ImageJ. Stack the images you want to analyze (Image/Stack/Images to Stack). Then, set the threshold of detection (Image/Apply/Threshold). Set it in a way that detects all specks (use the positive control), but does not detect the dimmer (non-activated) cells (use the negative control). Once done, apply this threshold to all images. After setting the threshold, go to Analyze/Analyze particles. The pixel size that should be entered depends on the magnification of the image and the size of the ASC specks. Make sure to visualize the results (Show: outlines) and check the box "Summary". Test different values for the pixel size and compare them to your own visual speck detection. When satisfied, apply the same thresholding algorithm and particle analysis to all images. In the summary window, you should then find the name of each image with the total count of specks per image.
5. Analysis of the DRAQ5 staining with ImageJ: Use the ICTN plugin and follow the instructions of the plugin.
6. Normalization: Divide the number of specks detected in **step 4** by the number of nuclei detected in **step 5**.

4 Notes

1. The quality of the HEK293T cells is of great importance for the success of retroviral transduction. They should be in a low passage number and never have been above 80 % confluency. Best results can be obtained if HEK293T cells used for virus production are cultured at around 50 % confluency.
2. If you would like to overexpress another protein in the same cell line, we recommend transducing the construct encoding the other protein before the one encoding ASC. We have found that this order results in more viable cells. Cells transduced with ASC often die when transduced with an additional construct. We have also successfully transduced HEK293T and THP-1, which are both human cell lines. Changes necessary to the protocol for THP-1 transduction: Use complete RPMI (RPMI 1640 + 10 % FBS, optional: antibiotics, e.g., 10 μg/mL Ciprofloxacin) as a plating medium for THP-1 cells. Do not

forget to pellet the THP-1 cells before removing the medium, as they are suspension cells. All subsequent selection steps also need to be carried out in complete RPMI.

3. We recommend using endotoxin-free maxi-prep DNA with a concentration above 1 μg/μL for generating retroviral particles.
4. We recommend using Luer lock syringes and filters to minimize the risk of leakage during the filtration of the virus supernatant. Make sure that you use 0.45 μm filters because 0.22 μm filters will block most virus particles and yield much lower virus titers.
5. Many inflammasome stimuli are toxic and need to be handled with care. Follow the manufacturer's instructions. As activators of NLRP3 the following conditions can be used (prime with 200 ng/mL LPS for 2–3 h before stimulation): 5 mM ATP for 30–60 min; 10 μM Nigericin for 60–90 min; 250–500 μg/mL Silica crystals for 4–6 h; 1 mM L-leucyl-L-leucine-methyl ester for 90–120 min; for AIM2: transfect 200 ng of poly(dA:dT) with 0.5 μL Lipofectamine 2000 per well (96-well) for 2–4 h.
6. Formaldehyde is toxic and carcinogenic. Wear protective equipment and dispose of formaldehyde according to the local regulations. Formaldehyde solutions need to be made up fresh every time, as formaldehyde oxidizes to formic acid over time. For convenience and to avoid the hazards during the generation of fresh formaldehyde, 16 % formaldehyde is available in ready-to-use ampules (Thermo Scientific).
7. The ImageJ protocol we present here is mostly manual and therefore only suitable for a limited number of pictures. Acapella Image Analysis Software (Perkin Elmer) or others are more suitable for high-throughput analysis.
8. It is advised to not use the outer wells of the plate, as they are more prone to evaporation. Thus, only use the middle 60 wells from 2 to 10 and from B to G.
9. Be very careful when removing the supernatants from the transfected HEK293T cells as they easily dislodge. Do not wash the cells.

   Since the retroviral vector construct is not only packaged, but also transiently expressed in HEK293T cells, they should be brightly fluorescent when investigated under the fluorescence microscope 24 h after transfection. If they are not, you should abort the virus production at this stage and troubleshoot: either the transfection was unsuccessful or you used the wrong constructs.
10. When we have low transduction efficiency, we combine both methods, performing first an antibiotic selection, followed by FACS with the stringent criteria outlined in **Note 11**.

11. If you have a high percentage of positive cells, you can apply additional criteria: Cells with weak fluorescence often form very small ASC specks, which can be hard to detect in automated imaging analysis. Highly fluorescent cells are prone to ASC speck formation even without stimulus and the signal-to-noise ratio between non-activated cells and ASC specks decreases. We usually divide the mCerulean positive cells into three gates: The weakest 15 %, the brightest 15–20 % and the medium-bright population in between. We investigate the populations under the fluorescent microscope and decide which population would be best. In most cases, the medium-bright cells give the best signal-to-noise ratios.
12. Perform a kill curve whenever you start working with a new cell line to determine the correct concentration of puromycin.
13. Make sure that the cells are 50 % confluent or less on the next day as selection is most efficient if the cells are 50 % or less confluent.
14. All cells in the non-transduced control should be dead. If not, continue selection and/or increase the concentration of puromycin until all cells in the non-transduced control are dead.
15. We use 0.5 cells/well because this leads to mostly single clones and very little wells with two or more clones. If you cannot generate any clones using this low concentration, we recommend increasing to 1–2 cells/well.
16. By this time, the immortalized macrophages form visible colonies that can easily be screened for. Depending on the cell type, this growth period needs to be extended (up to 14 days). If it takes longer than 7 days, add another 100 μL complete DMEM to the wells on day 7. Be sure to select clones as early as possible. Overgrowth can increase ASC speck formation and should be avoided even during the clonal selection procedure. However, do not be discouraged if some cells in a clonal colony show ASC speck formation—often, this is a temporary effect.
17. We recommend using single clones. The expression of ASC-mCerulean is detrimental for growth, so that in a mixed population, cells with a low expression of ASC-mCerulean will slowly overgrow the others and your cell line will change over time. Therefore, we suggest subcloning a successful clone for use in high-throughput screening approaches or if use over longer time periods is planned.
18. The more stimuli you test, the better your selection procedure. For NLRP3 stimuli, do not forget to prime the cells before stimulation (*see* **Note 5**). A good clone will respond to all known stimuli with at least 20 % ASC speck formation. There should be only one ASC speck per cell, which should be large and bright enough for image analysis.

19. Other counterstains are possible. In general, counterstaining of the nucleus gives best results, but might cause artifacts when stimulating the reporter cells with DNA or RNA. Membrane counterstains are not recommended because inflammasome activation leads to blebbing and fragmentation of the membranes, which make automated cell detection difficult. Also, the counterstaining should not involve any washing steps. Cells with an ASC speck will often detach and are easily lost during removal of the supernatants. This would lead to an underestimation of the activation of your stimulus.
20. Optional: Incubate at 37 °C for 30 min to improve the staining pattern of DRAQ5.

References

1. Masumoto J, Taniguchi S, Ayukawa K, Sarvotham H, Kishino T, Niikawa N, Hidaka E, Katsuyama T, Higuchi T, Sagara J (1999) ASC, a novel 22-kDa protein, aggregates during apoptosis of human promyelocytic leukemia HL-60 cells. J Biol Chem 274(48): 33835–33838
2. Schneider CA, Rasband WS, Eliceiri KW (2012) NIH Image to ImageJ: 25 years of image analysis. Nat Methods 9(7):671–675

Chapter 9

Immunoblotting for Active Caspase-1

Christopher Jakobs, Eva Bartok, Andrej Kubarenko, Franz Bauernfeind, and Veit Hornung

Abstract

Immunoblotting for caspase-1 is the gold-standard method of detecting inflammasome activation. In contrast to IL-1β-based readouts, it can be used in an experimental setup independent of de novo gene expression. Here, we present protocols for the preparation and precipitation of supernatant samples containing activated caspase-1 as well as protocols for polyacrylamide gel electrophoresis (PAGE) and protein immunoblotting.

Key words Caspase-1, Immunoblot, Western blot, Inflammasome

1 Introduction

A widely used readout for inflammasome and thus caspase-1 activation is the measurement of IL-1β release by ELISA, or the detection of IL-1β cleavage by immunoblotting. The major disadvantage of this approach is that assessment of mature cleaved IL-1β requires the induction of pro-IL-1β expression by a pro-inflammatory priming signal, such as a TLR ligand. However, priming can critically interfere with the subject under study, as various inflammasome complexes also require a pro-inflammatory priming signal [1]. As such, assessing caspase-1 activation directly remains the gold standard readout of inflammasome activation.

Since the proteolytic activation of caspase-1 via the inflammasome is the central step regulating its own proteolytic activity, detection of its subunits p20 or p10 is a good surrogate marker for caspase-1 activity. Currently, the standard technique of assessing caspase-1 cleavage is SDS-PAGE electrophoresis followed by immunoblotting. The activated caspase-1 is blotted from cell

---

Christopher Jakobs and Eva Bartok contributed equally to this work.

Christine M. De Nardo and Eicke Latz (eds.), *The Inflammasome: Methods and Protocols*, Methods in Molecular Biology, vol. 1040, DOI 10.1007/978-1-62703-523-1_9, © Springer Science+Business Media New York 2013

supernatants after protein precipitation whereas the uncleaved pro-form can be detected directly from cell lysates.

Detecting the caspase-1 subunits p20 or p10 by immunoblot presents certain technical difficulties, particularly with the choice of protein precipitation, protein separation and transfer protocols. Here, we describe several methods and provide specific recommendations for reliable analysis of caspase-1 cleavage by immunoblotting. To illustrate our approach, we use a murine immortalized macrophage cell line stimulated with defined NLRP3 activators (LPS priming followed by stimulation with ATP or Nigercin) [2]. However, this protocol is also amenable to other species, cell types, and of course other stimuli.

2 Materials

Prepare all reagents with ultrapure water and store them at room temperature (RT) unless otherwise specified. Please follow all waste disposal guidelines when disposing of reagents.

2.1 Cell Stimulation

1. 12-well tissue culture (TC) plates.
2. Immortalized murine macrophages (iMØ).
3. Cell culture medium: DMEM supplemented with 1 mM sodium pyruvate and 10 μg/mL ciprofloxacin (*see* **Note 1**). Store at 4 °C.
4. ATP: 500 mM Disodium-ATP dissolved in 500 mM NaOH. Store at −20 °C, and avoid freeze/thaw cycles.
5. Nigericin: 6.5 mM Nigericin stored in ethanol at −20 °C.
6. LPS: 5 mg/mL.
7. 2× SDS sample buffer: 150 mM Tris–HCl, pH 6.8, 200 mM DTT, 4 % SDS, 0.02 % bromophenol blue, 20 % glycerol. Store at −20 °C.

2.2 Protein Precipitation

2.2.1 Methanol–Chloroform Precipitation [3]

1. Methanol.
2. Chloroform.
3. SDS sample buffer (*see* **item** 7 of Subheading 2.1).
4. 1.5 mL tubes.

2.2.2 Trichloroacetate Precipitation [4]

1. 20 % Trichloroacetate (TCA).
2. Ice-cold Acetone.
3. 1 M Tris buffer, pH 6.8.
4. SDS sample buffer (*see* **item** 7 of Subheading 2.1).
5. Centrifuge at 4 °C.

2.3 Gel Electrophoresis

1. Ammonium persulfate (APS): 10 % solution in H_2O. Store at 4 °C (*see* **Note 2**).
2. *N*,*N*,*N*′,*N*′-tetramethylethylenediamine (TEMED). Store at 4–8 °C.
3. Propanol.
4. 8 cm (W) × 7.3 cm (H) mini gel casting system with plates and combs for a 1.5 mm gel (e.g., Bio-Rad Mini-PROTEAN 3 electrophoresis system).
5. Vertical gel electrophoresis and transfer tank system (e.g., Bio-Rad Mini-PROTEAN 3 electrophoresis system).

2.3.1 Tris-Glycine PAGE

1. 30 % Acrylamide/Bisacrylamide solution (37.5:1) (e.g., Carl Roth: Rotiphorese Gel 30, 3029.X). Store at 4–8 °C.
2. Sodium dodecyl sulfate (SDS): 10 % solution in H_2O (*see* **Note 3**).
3. 1 M Tris–HCl, pH 6.8.
4. 1.5 M Tris–HCl, pH 8.8.
5. 10× SDS Running Buffer: 250 mM Tris, 1.92 M Glycine, 1 % SDS in H_2O. For 1 L: 30.2 g Tris, 144 g Glycine, 10 g SDS and fill up to 1 L with H_2O.

2.3.2 Bis–Tris PAGE

1. 30 % Acrylamide Solution. Store at 4–8 °C.
2. 2 % Bisacrylamide Solution. Store at 4–8 °C.
3. 3.5× Bis–Tris Gel Buffer: 1.25 M Bis–Tris HCl, pH 6.6. To make: 26.16 g Bis–Tris made up to 100 mL with H_2O. Adjust pH to 6.6 with HCl.
4. 10× MES-SDS Running Buffer: 500 mM MES, 500 mM Tris–HCl, 1 % SDS and 10 mM EDTA. To make: 97.6 g MES, 60.58 g Tris–HCl, 10 g SDS made up to 1 L with H_2O (*see* **Note 4**).
5. 200× Running Buffer Reducing Agent: 1 M Na-Bisulfite in H_2O. For 50 mL: Dissolve 9.506 g Na-Bisulfite in 50 mL H_2O. Store at 4 °C.

2.4 Immunoblotting Reagents

1. 0.2 μm Nitrocellulose Membrane (*see* **Note 5**).
2. General-purpose blotting paper or Whatman paper.
3. 1 large measuring cylinder.
4. 1 small plastic container (e.g., the cover of a pipette tip box) for each blot.
5. Wet gel transfer system (e.g., Bio-Rad Protean) and an ice block which fits into the chamber.
6. 10× Tris–Glycine running buffer: 250 mM Tris, 1.92 M Glycine in H_2O. For 1 L: 30.2 g Tris, 144 g Glycine, fill to 1 L with H_2O.
7. 96 % Ethanol, denatured is acceptable for most systems (*see* **Note 6**).

8. Nonfat milk powder to make blocking solution.
9. Tween 20 %.
10. 10× PBS: For 1 L Add 80 g NaCl, 2 g KCl, 26.8 g $NA_2HPO_4{\cdot}7H_2O$, 2.4 g KH_2PO_4 and fill to a final volume of 1 L with H_2O. Adjust pH to 7.4 with HCl.
11. PBS–Tween Solution: 100 mL 10× PBS, 900 mL H_2O, 500 μL 20 % Tween (*see* **Note 7**).

2.5 Antibodies and Developing Tools

1. Primary antibodies detecting murine caspase-1 (*see* **Note 8**).
   - sc-514 (Santa Cruz): Rabbit polyclonal IgG anti p10 subunit, provided at 200 μg/ mL.
   - Casper-1 (Adipogen): Mouse mAb IgG1 anti p20 subunit, provided at 1,000 μg/mL.
2. Secondary antibodies (*see* **Note 9**).
   - sc-2004 (Santa Cruz): goat anti-rabbit IgG-HRP provided at 400 μg/mL.
   - sc-2005 (Santa Cruz): goat anti-mouse IgG-HRP provided at 400 μg/mL.
3. 50 mL conical tubes.
4. ECL Substrate kit.
5. Photographic films.
6. Film developing cassette.
7. Developing machine (*see* **Note 10**).

3 Methods

3.1 Cell Stimulation

1. Seed 5×10^5 immortalized murine macrophages [2] in 12-well plates in culture medium (500 μL/well).
2. Incubate cells overnight.
3. Stimulate the cells as follows.
   - Control: 6 h exposure to 200 ng/mL LPS.
   - ATP: 4 h priming with LPS, then stimulation with 5 mM ATP for 2 h.
   - Nigericin: 4 h priming with LPS, then stimulation with 650 μM Nigericin for 2 h.

After harvesting the supernatants, the remaining cells were washed with PBS and then directly lysed in 150 μL 1× SDS sample buffer per well. For each condition, the supernatants and lysates were pooled separately to minimize loading differences while comparing protocols. Pooled supernatants were centrifuged at 1,000 ×*g* for 5 min to remove cellular debris. Both supernatants and lysates were stored at −20 °C until use.

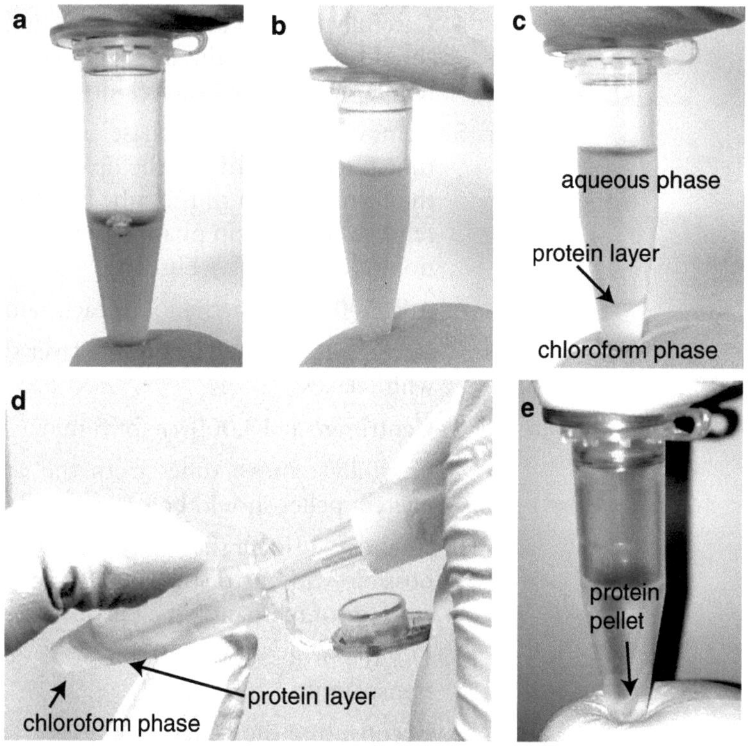

Fig. 1 Chloroform–methanol precipitation. (**a**) 1.5 mL tube containing 500 μL of supernatant from cultured macrophages. (**b**) 500 μL supernatant is mixed with 500 μL of methanol and 125 μL of chloroform and vortexed. (**c**) After centrifugation a protein layer between the aqueous and organic phase is visible. (**d**) The aqueous phase is removed without disturbing the protein layer using a suction device. (**e**) After centrifugation with an additional 500 μL of methanol a protein pellet forms at the site of maximal centrifugation force

3.2 Sample Precipitation

In order to detect cleaved caspase-1 in the supernatant, it is advisable to concentrate the sample via precipitation. From one 12-well TC plate, use 500 μL of supernatant for precipitation (*see* **Note 11**). The samples can be used fresh or thawed after storage at −20 °C.

Although most standard protein precipitation protocols work sufficiently well (*data not shown*), we found that methanol–chloroform precipitation and TCA-precipitation provide the most reliable results (*see* **Note 12**). Thus, we have only included adapted versions of these two protocols in detail. All centrifugation steps are carried out at room temperature unless otherwise indicated.

3.2.1 Methanol/Choloroform Precipitation

1. Add 500 μL methanol and 125 μL chloroform to 500 μL supernatant.
2. Vortex the sample vigorously for at least 30 s. The sample should be a homogenous, milky white suspension (*see* Fig. 1b).
3. Centrifuge the sample at 13,000 × *g* for 5 min.

4. Carefully remove the tubes from the centrifuge. Three phases should be visible: the upper aqueous/methanol phase, a protein layer, and the lower chloroform phase (*see* Fig. 1c).
5. Remove the aqueous phase using a suction device with the tube tilted forward and the tip of the suction device directed to the front portion of the tube (*see* **Note 13**). Be careful not to remove the protein phase or protein clinging to the back portion of the tube (*see* Fig. 1d).
6. Add 500 μL of methanol to each sample.
7. Vortex vigorously. The protein layer should break up into small white flakes.
8. Centrifuge at 13,000 × *g* for 5 min.
9. Carefully remove tubes from the centrifuge. A small, white protein pellet should be visible in all samples (*see* Fig. 1e).
10. As above, using a suction device, remove as much methanol as possible without disturbing the pellet. Try to leave the same amount of methanol in all probes (*see* **Note 13**).
11. Dry the probes for 5–10 min at 55 °C. The pellet will become transparent.
12. Reconstitute the pellet in 40 μL of 1× SDS sample buffer. Do not forget to scrape the back wall of the tube when resuspending the pellet.
13. Vortex the samples vigorously for 1 min.
14. Centrifuge for a few seconds at 13,000 × *g* to settle the contents of the tube.
15. Incubate the samples at 95 °C for 5 min.
16. Vortex the samples vigorously for 1 min.
17. Cool the samples by placing them on ice.
18. Quickly spin tubes (e.g., 30 s centrifugation at 10,000 × *g*) to collect the samples at the bottom of the tubes.

3.2.2 Trichloroacetate Precipitation

1. Add 500 μL of 20 % TCA to 500 μL of supernatant (1:1).
2. Incubate on ice for 30 min.
3. Centrifuge at 12,000 × *g*, 4 °C for 15 min.
4. Wash the samples 3×.
   - Carefully remove the supernatant
   - Add 500 μL ice cold acetone
   - Centrifuge at 12, 000 × *g*, 4 °C for 5 min
5. Remove supernatant and air-dry pellet.
6. Reconstitute the pellet in 40 μL SDS sample buffer (*see* **Note 14**).
7. Vortex and boil samples as in **steps 14–18** of Subheading 3.2.1.

3.3 Polyacrylamide Gel (PAGE) Electrophoresis

Bis–Tris PAGE offers a superior resolution for the separation of low-molecular weight proteins. This is achieved by the operating pH of 7.0 (as opposed to a pH of 9.5 of Tris–Glycine gels), which protects protein samples from deamination and alkylation and therefore increases protein stability during electrophoresis [5] (*see* **Note 15**). Moreover, Bis–Tris gels have the advantage of shorter running times and a longer shelf life. Nevertheless, we have also included a standard Tris–Glycine PAGE protocol because it is a good alternative for labs where Bis–Tris gels are not available. To blot the 10 kDa and 20 kDa activated caspase-1 from supernatants, we recommend using a 10 % Bis–Tris PAGE or a 15 % Tris–Glycine PAGE. For 40 kDa pro-caspase-1 electrophoresis, we recommend using an 8 % Bis–Tris gel or a 12 % Tris–Glycine PAGE.

3.3.1 Bis–Tris-PAGE Electrophoresis

1. Prepare 8 mL of 10 % resolving gel by mixing 656 μL H_2O, 2.5 mL 30 % acrylamide, 2.5 mL 2 % Bisacrylamide, 2.286 mL 3.5× Bis–Tris gel buffer, 50 μL APS and 8 μL TEMED (*see* **Note 16**). Vortex the mixture.
2. Cast gel in the assembled mini-gel chamber with 1.5 mm spacers. Layer the gel with 500 μL isopropanol. The gel will need around 10 min to harden.
3. Prepare 4 mL of 4 % stacking gel by mixing 1.83 mL H_2O, 500 μL 30 % Acrylamide, 500 μL 2 % Bisacrylamide, 1.14 mL 3.5× Bis–Tris gel buffer, 25 μL APS and 4 μL TEMED (*see* **Note 16**).
4. Remove isopropanol and layer the stacking gel over the resolving gel. Immediately add the 1.5 mm, 15-well comb to the gel without introducing air bubbles.
5. Prepare 1× MES-SDS running buffer from 10× MES-SDS stock solution (*see* **item 4** of Subheading 2.3.2) and add 0.005 % of 200× reducing agent immediately before use.
6. When the gel has hardened, remove the comb and assemble the electrophoresis chamber. Pour the running buffer in both chambers and load the samples.
7. Run the gel for ~1 h at 100 V.

3.3.2 Tris–Glycine-PAGE Electrophoresis

1. Prepare 8 mL of 15 % resolving gel by mixing 1.84 mL H_2O, 4 mL 30 % Acrylamide/Bisacrylamide solution, 2 mL 1.5 M Tris–HCl pH 8.8, 80 μL SDS solution, 80 μL APS solution and 8 μL TEMED (*see* **Note 16**).
2. Cast gel in the assembled mini-gel chamber with 1.5 mm spacers. Layer the gel with 500 μL isopropanol. The gel will need around 10 min to harden.
3. Prepare 4 mL of 5 % stacking gel by mixing 2.72 mL H_2O, 0.68 mL 30 % Acrylamide/Bisacrylamide solution, 0.52 mL 1 M Tris–HCl pH 6.8, 40 μL SDS solution, 40 μL APS solution and 4 μL TEMED (*see* **Note 16**).

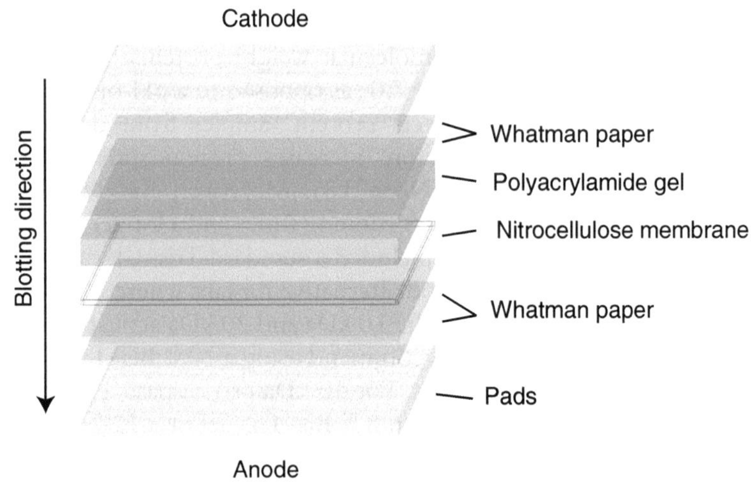

Fig. 2 Assembly for blotting. A schematic view of the assembly setup for the blotting procedure is shown

4. Remove isopropanol and layer the stacking gel over the resolving gel. Immediately add the 1.5 mm, 15-well comb to the gel without introducing air bubbles.
5. Prepare 1 L of 1× SDS running buffer from the 10× stock solution (*see* **item 5** of Subheading 2.3.1).
6. When the gel has hardened, remove the comb and assemble the electrophoresis chamber. Pour the running buffer in both chambers and load the samples.
7. Run the gel for ~2 h at 100 V.

3.4 Gel Transfer

We recommend using a short transfer (30 min) for the 15 % Tris–Glycine gel, especially if the small p10 subunit of caspase-1 is to be detected. For the 10 % Bis–Tris gel, we recommend a transfer time of 45–60 min (*see* Fig. 3).

1. Following electrophoresis, pry the glass plates open so that the gel will remain on one of the glass plates. Remove the stacking gel.
2. For both gel types, prepare the transfer buffer by mixing 200 mL 10× Tris–Glycine, 400 mL Ethanol and 1,400 mL H_2O (*see* **Note 6**).
3. Incubate the gel for 10 min in the transfer buffer.
4. Prepare two pieces of Whatman paper per gel.
5. Prepare a piece of nitrocellulose membrane large enough to cover the lanes of interest and small enough to fit into the transfer cassette.
6. Assemble the transfer as displayed in Fig. 2.

7. Set up the transfer chamber and pour at least 500 mL of transfer buffer.
8. Place the cassette and two sponges in the buffer.
9. Carefully open the gel mold and assemble the gel cassette as in Fig. 2 (*see* **Note 17**).
10. Place the cassette in the transfer chamber and fill the chamber with transfer buffer. For cooling during transfer add a chilled cool pack into the chamber or place the chamber in a cooling box surrounded by ice.
11. Transfer for 30 min (15 % Tris–Glycine gel) or 45 min (10 % Bis–Tris gel) at 100 V.

3.5 Incubation with Antibodies and Blot Exposure

We have found that the commercially available antibodies sc-514 (Santa Cruz) and Casper-1 (Adipogen) work well for immunoblotting of the p10 and p20 subunit respectively (*see* Fig. 3 and **Note 8**). Wash and block the membranes separately using one small plastic container for each blot.

1. Dilute 1:10 of a sufficient quantity of 10× PBS in water. You will need 50–100 mL per blot and wash cycle.
2. Wash each membrane in 1× PBS (without Tween) 3× for 5 min.
3. Prepare 3 % milk blocking solution. You will need ~25 mL for each blot. For 50 mL: mix 5 mL 10× PBS, 30 mL H_2O, and 1.5 g milk powder until the milk dissolves and then fill to 50 mL with H_2O (*see* **Note 18**).
4. Block membranes in 3 % milk blocking solution for 1 h at RT.
5. Wash the membranes 4× in PBS with 0.05 % Tween for 5 min.
6. Prepare a 1 % milk blocking solution for Antibody Incubation (as in **step 3** of Subheading 3.5), but use only 0.5 g milk powder for 50 mL blocking solution. You will need ~3 mL solution for each blot (*see* **Note 18**).
7. Add the primary antibody to the 1 % milk blocking solution with the appropriate dilution.
   - Sc-514: 1:200 (final concentration 1 μg/mL).
   - Casper-1: 1:1,000 (final concentration 1 μg/mL).
8. Incubate the membrane in a 50 mL conical tube with at least 3 mL of antibody solution on a rotator overnight at 4 °C.
9. Wash the membranes as indicated in **step 5** of Subheading 3.5.
10. Prepare 1 % milk blocking solution for Antibody Incubation: as in **step 3** of Subheading 3.5 but use only 0.5 g milk powder for 50 mL blocking solution. You will need approx. 3 mL solution for each blot (*see* **Note 18**).
11. Dilute the appropriate conjugated secondary antibody 1:2,000 in the 1 % milk blocking solution (*see* **Note 9**).

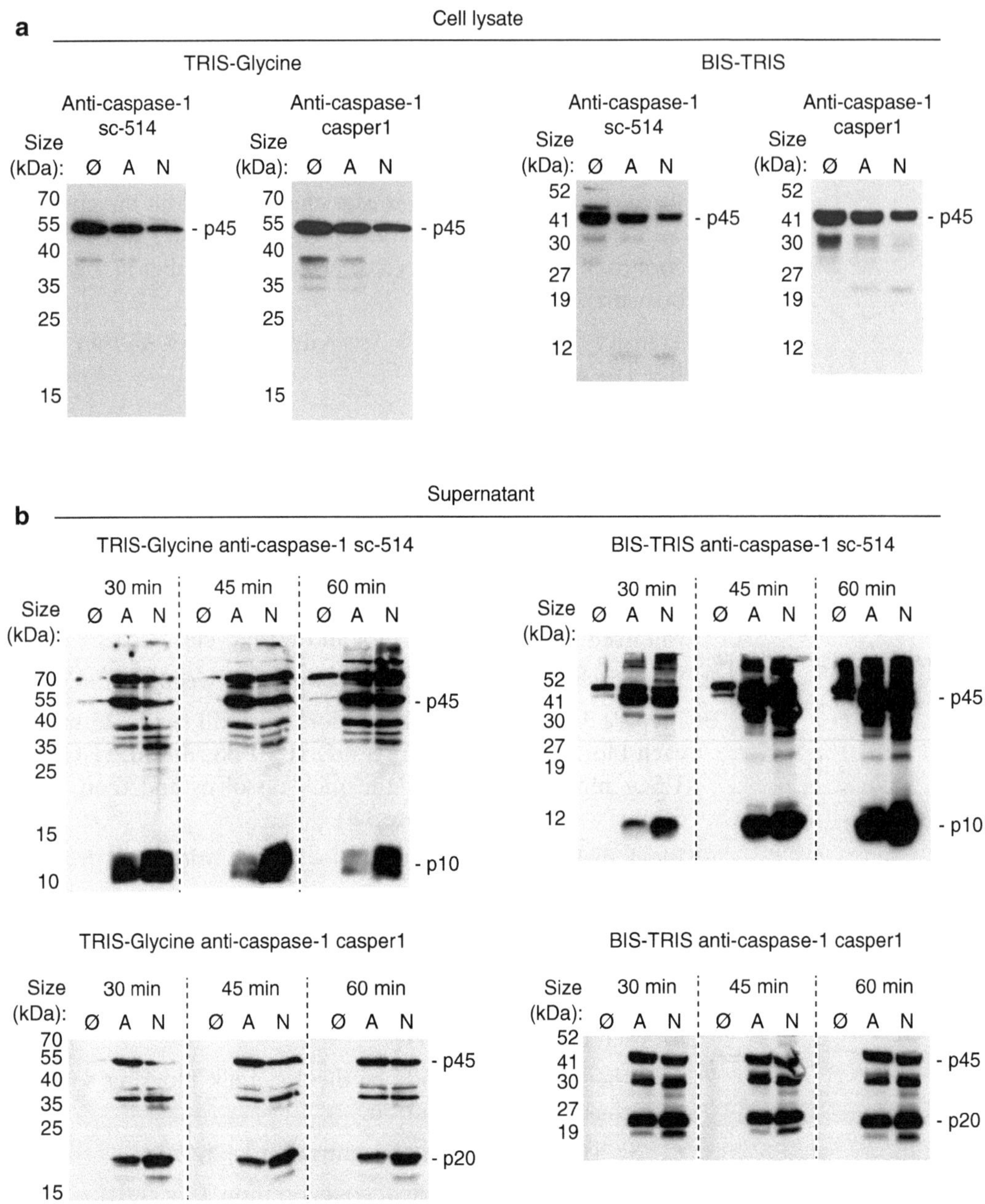

Fig. 3 Caspase-1 western blots. 5×10^5 LPS-primed macrophages were stimulated with either PBS (*Ø*) or ATP (*A*) or Nigericin (*N*) in 500 μL serum-free DMEM for 2 h. (**a**) After stimulation cells were lysed in Laemmli buffer and lysates were separated using Tris–Glycine or Bis–Tris gels. Following transfer, membranes were stained for caspase-1 by with anti-caspase-1 antibody sc-514 and subsequently stripped and re-probed with Casper-1. (**b**) Supernatants of stimulated cells were precipitated by chloroform–methanol and subsequently separated using Tris–Glycine or Bis–Tris gels. For each lane 40 % (16 μL) of a 500 μL supernatant precipitation was loaded. The gels were each divided into three parts and then blotted for 30′, 45′ or 60′ onto nitrocellulose. Following transfer, membranes were stained for caspase-1 by using anti-caspase-1 antibody sc-514 and subsequently stripped (*see* **Note 21**) and re-probed using Casper-1

- For sc-514 use anti-rabbit-HRP.
- For Casper-1 use anti-mouse-HRP.

12. Incubate the membrane in a 50 mL conical tube with at least 3 mL conjugated antibody solution on a rotator for 90 min at RT. Keep the tube protected from light.
13. Wash the membranes as indicated in **step 5** of Subheading 3.5.
14. Briefly dry the membrane by dipping the edges on cellulose and apply sufficient ECL substrate (e.g., 2 mL per mini blot) onto the membrane and wait 30 s (*see* **Note 19**).
15. Drain superfluous ECL substrate from the membrane and place it between two layers of clear foil. Carefully scrub the foil to remove remnants of substrate and to remove air bubbles and put it into a film exposure cassette box.
16. Apply a film onto the membrane in a dark room.
17. The amount of time needed for the film exposure varies considerably. For the figures shown (*see* Fig. 3b), the film was exposed for approximately 10 s (*see* **Note 20**).

4 Notes

1. It is critical not to use FCS-supplemented medium if supernatants are to be precipitated. The excess protein will overload the gel during electrophoresis, resulting in streaking bands. We recommend using serum-free medium or serum-reduced medium.
2. A new stock should be made every month. Alternatively, aliquots can be kept at −20 °C.
3. It is usually necessary to mix overnight with a magnetic stirrer.
4. We recommend using MES-SDS buffer instead of the commonly used MOPS-SDS buffer, as it is better suited for small proteins such as the p10 and p20 subunits of caspase-1 [6].
5. It is also possible to use membranes with 0.45 μm pores, yet different transfer times might be required.
6. Denatured 96 % ethanol can be used if the membrane is to be developed by conventional indirect film exposure. For infrared imaging systems that directly detect the protein on the membrane (e.g., Li-Cor Odyssey), pure ethanol should be used to avoid interference.
7. Tween 20 % has a rather high viscosity. Thus, it is best to cut off the narrow portion of the pipette tip and not to flush the tip after releasing the Tween.

8. For immunoblotting of human caspase-1, we obtain good results using sc-515 (Santa Cruz) a rabbit polyclonal IgG antibody provided at 200 μg/mL.
9. These are HRP-conjugated antibodies for ECL substrate and a conventional developer. There are suitable conjugated antibodies for infrared imaging available on the market, as well. Please also consult the manual of the imaging system that will be used.
10. We employ conventional photographic film exposure in a cassette and a developing machine, yet other luminescence-based or fluorescence-based detection systems are also well suited.
11. With more experience, it is possible to use less supernatant, e.g., 100–200 μL. However, such changes should only be made after the immunoblotting protocol is well established.
12. Even if acetone and ethanol precipitation are both effective, they are less suitable for the precipitation of samples containing large quantities of ATP. The ATP is precipitated along with the protein and causes running difficulties.
13. The centrifuge will distribute some protein on the back wall of the tube, so be careful to always face all samples in the same direction, i.e., with the hinge of the tube to the back. When removing the methanol or TCA, be careful not to touch the back wall of the tube, so that the protein there will not be lost. Do not risk losing the pellet by taking out too much methanol, but try to only leave a small, uniform quantity, because the samples should all need the same amount of time to dry.
14. Bromophenol blue turns yellow in an acidic milieu (pH 3). If the sample turns yellowish, you can adjust the pH with a small volume of 1 M Tris, pH 6.8.
15. The neutral pH of the running front alters the migration behavior of some prestained markers, which are usually calibrated to run in Tris–Glycine gels (pH = 9.5). It is recommended to check the compatibility of the marker prior to electrophoresis or to use an unstained marker for correct determination of protein size. In our case, we used the PageRuler™ prestained marker (Fermentas) and calibrated it using the PageRuler™ Unstained Broad Range Protein Ladder (Fermentas).
16. Adding APS and TEMED will start gel polymerization and should only be done immediately before the gel is cast.
17. This is the description of a wet transfer. It is also possible to put together the cassette in a semidry fashion.
18. It is best to make a fresh milk blocking solution on the day of use.
19. We use the Pierce ECL Western Blotting Substrate or our self-made substrate as described [7] as an HRP-substrate. However, the substrate needed depends on the conjugated antibody and

general developing technique. Please also consult the manual of the imaging system that will be used.

20. The amount of exposure time depends on many variables and thus it is difficult to give a general recommendation. Although we only needed 10 s exposure for the blots shown, it may even be necessary to expose the film overnight. In the beginning we recommend trying out different exposure times in the range of 10 s to 30 min.

21. Stripping was performed by membrane incubation for 30 min at 55 °C in a 50 mL conical tube completely filled with stripping buffer (2 % SDS, 62.5 mM Tris–HCl pH 6.8, 0.8 vol.% of β-Mercaptoethanol in H_2O). During the incubation the tube was carefully inverted ten times. Subsequently, the membrane was thoroughly washed (*see* **step 2** of Subheading 3.5) and the membrane was furthermore processed starting from **step 3** of Subheading 3.5.

References

1. Bauernfeind FG, Horvath G, Stutz A, Alnemri ES, MacDonald K, Speert D, Fernandes-Alnemri T, Wu J, Monks BG, Fitzgerald KA, Hornung V, Latz E (2009) Cutting edge: NF-kappaB activating pattern recognition and cytokine receptors license NLRP3 inflammasome activation by regulating NLRP3 expression. J Immunol 183(2):787–791. doi:10.4049/jimmunol.0901363
2. Hornung V, Bauernfeind F, Halle A, Samstad EO, Kono H, Rock KL, Fitzgerald KA, Latz E (2008) Silica crystals and aluminum salts activate the NALP3 inflammasome through phagosomal destabilization. Nat Immunol 9(8):847–856. doi:10.1038/ni.1631
3. Wessel D, Flugge UI (1984) A method for the quantitative recovery of protein in dilute solution in the presence of detergents and lipids. Anal Biochem 138(1):141–143
4. Rajalingam D, Loftis C, Xu JJ, Kumar TK (2009) Trichloroacetic acid-induced protein precipitation involves the reversible association of a stable partially structured intermediate. Protein Sci 18(5):980–993. doi:10.1002/pro.108
5. Moos M Jr, Nguyen NY, Liu TY (1988) Reproducible high yield sequencing of proteins electrophoretically separated and transferred to an inert support. J Biol Chem 263(13): 6005–6008
6. Good NE, Winget GD, Winter W, Connolly TN, Izawa S, Singh RM (1966) Hydrogen ion buffers for biological research. Biochemistry 5(2):467–477
7. Haan C, Behrmann I (2007) A cost effective non-commercial ECL-solution for Western blot detections yielding strong signals and low background. J Immunol Methods 318(1–2): 11–19. doi:10.1016/j.jim.2006.07.027

Chapter 10

Inflammasome Activation and Inhibition in Primary Murine Bone Marrow-Derived Cells, and Assays for IL-1α, IL-1β, and Caspase-1

Katharina S. Schneider, Christina J. Thomas, and Olaf Groß

Abstract

Through its ability to control the proteolytic maturation and secretion of interleukin-1 family cytokines, the inflammasome occupies a central role in the activation of inflammation and also influences the shaping of adaptive immunity. Since it affects a multitude of different immune responses from autoinflammatory diseases to host defense, vaccine efficacy, and even cancer, it has become of interest to many researchers. Here, we describe a straightforward method for inflammasome assays in primary murine bone marrow-derived myeloid cells. The protocol encompasses cell handling, inflammasome activation and inhibition, as well as the detection of IL-1β, caspase-1, and IL-1α by ELISA and Western blot.

Key words Inflammasome, NLRP3, Caspase-1, Interleukin-1β, Interleukin-1α

1 Introduction

Autocatalytic cleavage of caspase-1, subsequent cleavage of IL-1β, and secretion of both factors are hallmarks of inflammasome activation [1]. IL-1α, while not a caspase-1 substrate, can also be secreted alongside IL-1β in an inflammasome-dependent manner [2–5]. Since both cytokines activate the same receptor [6], IL-1α might contribute to the beneficial or adverse effects of inflammasome activity. Thus, the direct comparison of IL-1α and IL-1β secretion in vitro and in vivo can be informative. Both IL-1 cytokines are encoded on the same genomic locus, and share a very similar protein structure [7]. A priming signal is required to induce their expression in innate immune cells of the myeloid lineage [3]. As both IL-1α and IL-1β lack a signal peptide, they are not released by the default, ER-Golgi-dependent secretion pathway, but are instead retained in the cytoplasm until a second signal, i.e., inflammasome activation, prompts their secretion [8].

Christine M. De Nardo and Eicke Latz (eds.), *The Inflammasome: Methods and Protocols*, Methods in Molecular Biology, vol. 1040, DOI 10.1007/978-1-62703-523-1_10, © Springer Science+Business Media New York 2013

Caspase-1-mediated removal of the pro-domain of IL-1β is required for its binding to the IL-1 receptor. In contrast to IL-1β, IL-1α is not a substrate of caspase-1, which could explain why IL-1α has been largely overlooked in the inflammasome field. Instead, the pro-domain of IL-1α is proteolytically processed by calpains [9, 10]. In the context of inflammasome activation, calpain activity can be induced in an inflammasome-dependent manner, indicating that the inflammasome can control IL-1α cleavage, albeit indirectly [3]. However, IL-1α cleavage is not required for receptor activation [7]. Therefore, any manner of IL-1α release (i.e., active release after inflammasome activation, or passive release after cell injury) is sufficient for receptor activation on neighboring cells. This has encouraged the view that IL-1α is a danger signal that passively leaks from the cytoplasm upon cell death [6], and IL-1α release has even been used as a readout for cell death. However, since the inflammasome can control its release, IL-1α appears to have a dual function as both a passively released danger signal and an actively secreted cytokine in the context of inflammasome activity.

Since its discovery over 10 years ago, the inflammasome has become one of the fastest growing areas of research in immunology [11, 12]. Initial studies were performed using cell lines such as THP1 monocytic cells and J774 macrophage-like cells. Aside from these two cell lines, most myeloid cell lines are hyporesponsive to inflammasome activators, presumably due to the low expression of certain inflammasome components. The generation of knockout mice deficient in the key inflammasome components caspase-1, Asc, and Nlrp3, and the use of these mice in in vivo disease models, has allowed identification of central roles of the different types of inflammasomes in host defense and in the pathogenesis of a multitude of diseases [1, 13–21]. In addition, primary myeloid cells derived from these mice are widely used in in vitro inflammasome research and have largely replaced the use of cell lines. Murine peritoneal macrophages, bone marrow neutrophils, bone marrow-derived macrophages (BMDMs), and bone marrow-derived dendritic cells (BMDCs) are all responsive to inflammasome activators.

This protocol describes our standard technique for priming, stimulating and inhibiting the inflammasome in primary murine BMDCs, as well as the measurement of IL-1α, IL-1β, and caspase-1 cleavage and secretion by ELISA and Western blot. It is optimized for the use of BMDMs and BMDCs, which are most commonly used in the inflammasome field because they have the distinct advantage that they can be generated in large quantities. We favor BMDCs due to the higher amounts of IL-1 produced [22]. Protocols for the differentiation of BMDCs and BMDMs have been reported [22–24]. An advantage of using primary cells is that inflammasome-deficient mouse lines are readily available commercially (e.g., caspase-1−/− mice [5] from The Jackson

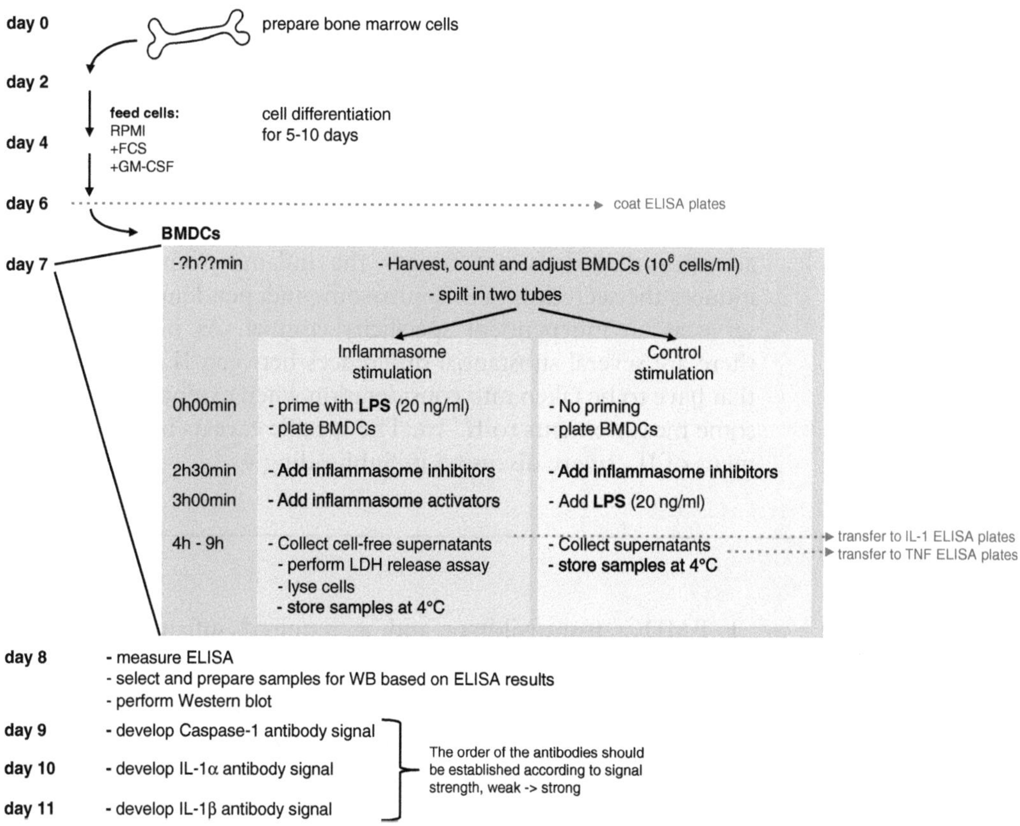

Fig. 1 Flowchart summarizing the method. *WB* Western blot

Laboratory) or can be obtained from the originating laboratories. It is therefore often easier to acquire inflammasome knockout mice (or bone marrow thereof) and to generate primary knockout cells, as compared to performing RNA interference in cell lines. An alternative to or an extension of this method is the generation of inflammasome deficient cell lines (and wildtype controls) through immortalization of primary mouse macrophages, for example using the J2 virus [25] (e.g., produced by AMJ2 cells). This can be especially helpful for researchers who want to use knockout cells, but want to avoid the expense of housing mice that are only used for in vitro studies.

The NLRP3 inflammasome is the most widely studied member of the inflammasome family, and responds to a multitude of readily available activators, making it the focus of this protocol. However, this protocol can also be applied to studies of the NLRP1, NLRC4, AIM2, and RIG-I inflammasomes. We propose a two-step process in which all supernatant samples are first measured by ELISA for the secretion of IL-1α and IL-1β. In a second step, both supernatants and cell lysates of selected samples are subjected to analysis by Western blot for IL-1α, IL-1β, and caspase-1 (*see* Fig. 1). ELISA allows more precise quantification, while

Western blot provides information about protein cleavage. Figures 2 and 4 show typical Western blots probed for these proteins. In addition, the protocol contains two different control measurements. We suggest combining the measurement of IL-1α with a quantification of cell death through a lactose dehydrogenase (LDH) release assay, in order to allow differentiation of the two possible causes of IL-1α release. Furthermore, stimulation with an activator that does not trigger the inflammasome but instead induces the secretion of inflammasome-independent cytokines can serve as an independent specificity control. As outlined above, there are several substantial differences between IL-1α and IL-1β that have to be taken into consideration when extending inflammasome measurements to IL-1α. The specific caveats in the measurement of IL-1α are discussed in Subheading 4.

2 Materials

2.1 Inflammasome Activation

1. BMDCs from wildtype and, as required, inflammasome-deficient mouse lines, day 5–10 of differentiation (*see* **Note 1**).
2. BMDC medium: RPMI1640 containing GlutaMAX™ (Life Technologies) supplemented with 10 % FCS, 100 units/mL penicillin/streptomycin, 50 μM β-mercaptoethanol, 10 mM HEPES, and 20 ng/mL recombinant murine GM-CSF (*see* **Note 2**).
3. 50 mL tubes.
4. Hank's Balanced Salt Solution containing 5 mM EDTA (HBSS/EDTA).
5. Trypan blue.
6. Hemocytometer.
7. 12-tip multichannel pipette with a volume range of at least 50–200 μL per tip.
8. 10 mL serological pipettes.
9. 1.5 mL tubes.
10. 95 °C heat block.
11. Cell culture incubator at 37 °C and 5 % CO_2.
12. 96-well flat-bottom cell culture plates.
13. Inflammasome activators, e.g.,
    - 1 M ATP stock (Sigma). Dissolve 1 g ATP in 1.8 mL H_2O. Store at −20 °C.
      Dilution factor 1:33.3 for a 6× working stock and 1:200 final = 5 mM.

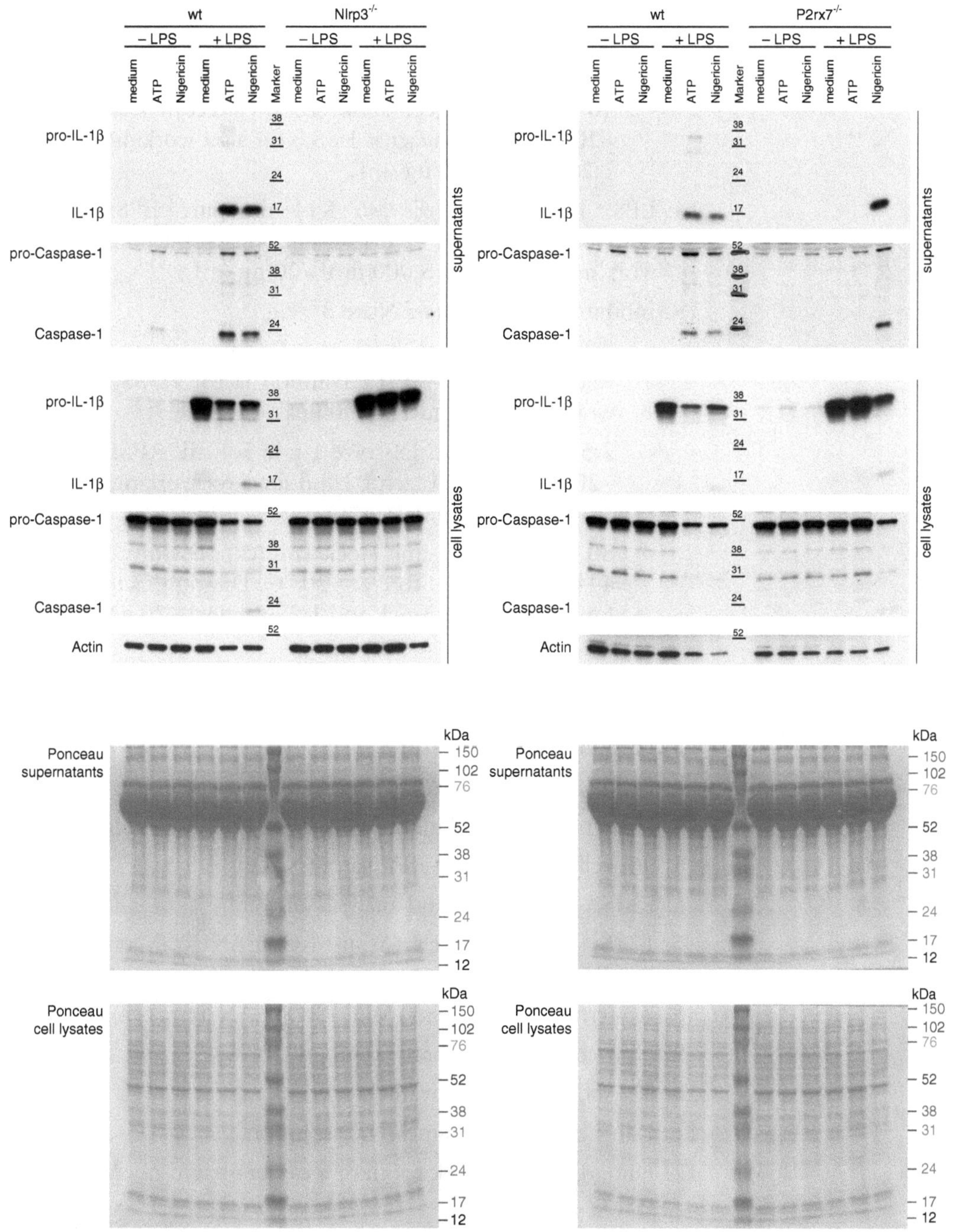

Fig. 2 Representative Western blots of cell-free supernatants and cell lysates of murine BMDCs. BMDCs from wildtype (wt) or Nlrp3-deficient (Nlrp3−/−, *left half*) or ATP receptor P2rx7-deficient (P2rx7−/−, *right half*) mice were LPS-primed and inflammasome-stimulated as indicated. Upregulation and processing of IL-1β and caspase-1 were analyzed by Western blot (*upper part*). Actin is shown as a loading control for lysates. Ponceau stainings of immunoblots (*lower part*) serve as a reference

- 10 mM nigericin stock (Sigma). Dissolve 5 mg in 670 μL EtOH. Store at 4 °C. Dilution factor 1:333.3 for a 6× working stock and 1:2,000 final = 5 μM.
- 40 mg/mL Imject alum stock (Pierce). Store at 4 or −20 °C. Dilution factor 1:33.3 for a 6× working stock and 1:200 final = 200 μg/mL.

14. LPS: 100 μg/mL *E. coli* K14 ultrapure LPS in water (Invivogen). Store at −20 °C. Dilution factor 1:833.3 for a 6× working stock and 1:5,000 final = 20 ng/mL.
15. Inhibitors of choice (*see* **Note 3**), e.g.,
    - 0.1 M Ebselen stock (Enzo). Dissolve 10 mg in 365 μL water. Store at −20 °C. Dilution factor 1:333.3 for a 6× working stock and 1:2,000 final = 50 μM.
    - 2.5 M KCl stock. Dissolve 1 g in 5.4 mL RPMI. Store at −20 °C. Dilution factor 1:8.3 for a 6× working stock and 1:50 final = 50 mM.
    - 50 mM Z-VAD-fmk stock (Merck). Dissolve 10 mg in 441 μL DMSO. Store at −20 °C. Dilution factor 1:416.7 for a 6× working stock and 1:2,500 final = 20 μM.

2.2 Inflammasome Measurement

1. Cell culture centrifuge with swing-out buckets to hold 96-well plates.
2. Phosphate-buffered saline (PBS).
3. ELISA kits for IL-1α, IL-1β, and TNF.
4. LDH release assay kit (Promega).
5. 96-well MaxiSorb ELISA plates (Nunc).
6. ELISA plate reader and analysis software.
7. 12 % SDS-PAGE gels (Mini-PROTEAN TGX precast gels, Bio-Rad).
8. Western blot equipment (Mini-PROTEAN Tetra cell, Bio-Rad).
9. Nitrocellulose membrane 0.45 μm pores.
10. 5 % sodium azide in water.
11. Enhanced chemiluminescence solution.
12. Film, developer, dark room or a camera-based documentation system.
13. 3× SDS PAGE sample buffer: 187.5 mM Tris–HCl, pH 6.8, 6 % w/v SDS, 0.03 % w/v Phenol Red, 30 % w/v Glycerol. Make a 1× stock by diluting the 3× stock with distilled water.
14. 5× running buffer: 15 g/L Tris base, 72 g/L Glycine, 5 g/L SDS.
15. 1× blotting buffer: 2.5 g/L Tris base, 12 g/L Glycine, 15 % denatured ethanol.

16. 1× Ponceau staining solution: 0.05 % Ponceau, 3 % Trichloroacetic acid.
17. Wash buffer: PBS containing 0.1% Tween-20.
18. Blocking buffer: wash buffer containing 2% skim milk powder.
19. Antibodies, e.g.,
    - Polyclonal anti-mouse IL-1β (goat, AF-401, R&D Systems).
    - Monoclonal anti-mouse caspase-1 p20 (mouse, Casper-1, AdipoGen).
    - Monoclonal anti-mouse IL-1α (hamster, ALF-161, eBioscience).
    - HRP-conjugated anti-goat IgG (rabbit, #6026-05, Southern Biotech).
    - HRP-conjugated anti-mouse IgG (goat, #115-035-146, Jackson IR).
    - HRP-conjugated anti-hamster IgG (mouse, #554012, BD Biosciences).

3 Methods

3.1 Stimulation of BMDCs

1. Plan the experiment and calculate the required number/volume of cells per mouse/genotype (*see* **Note 4**). Especially when inhibitors are used, it is crucial to perform a control experiment (*see* **Note 5**).
2. Harvest BMDCs between day 5 and day 10 of culture, by first transferring the suspension cells into 50 mL tubes (*see* **Note 6**).
3. Add 10 mL of HBSS/EDTA to the adherent cells remaining on the cell culture plate(s) and incubate for 10 min at 37 °C, to dislodge these cells.
4. Rinse the plate(s) thoroughly with the HBSS/EDTA by pipetting up and down with a serological pipette and add the collected cells to the respective 50 mL tube containing the suspension cells of the same plate/genotype.
5. Centrifuge the cells at 300–350 × g for 5–7 min and discard the supernatant.
6. Gently resuspended the cells in 10 mL of fresh BMDC medium using a serological pipette. Do not vortex (*see* **Note 7**).
7. Determine the total number of cells using trypan blue and a hemocytometer and adjust them to 10^6 per mL with fresh BMDC medium (*see* **Note 8**).
8. Split the cells into two tubes containing the volumes for inflammasome stimulation and control experiment (calculated in **step 1** of Subheading 3.1).

9. For inflammasome stimulation: Prime cells with 20 ng/mL LPS (1:5,000 of 100 μg/mL stock), and mix by inversion. Leave the cells intended for the LPS-induced TNF production assay (used to test for toxicity and off-target effects of the inhibitors) unprimed.
10. Plate 200 μL of both primed and unprimed cells per well in separate 96-well flat-bottom plates to perform triplicates of each condition.
11. Incubate the plates for 2.5 h at 37 °C to allow cellular priming to occur.
12. Prepare 6× stocks of inhibitors using dilution information in **item 15** of Subheading 2.1 (*see* **Note 9**).
13. Add 50 μL of the 6× inhibitor stocks (or medium) to the LPS-primed cells (to be treated with inflammasome activators) and unprimed cells (to be treated with LPS) and incubate for 30 min at 37 °C (*see* **Note 10**).
14. Prepare 6× concentrated inflammasome activators and 6× concentrated LPS (*see* **Note 11**).
15. Add 50 μL of 6× inflammasome activator to the primed cells and 50 μL of 6× LPS to the unprimed cells.
16. Incubate for 1–6 h (for inflammasome activators) or 4–6 h (for LPS) at 37 °C (*see* **Note 12**).
17. Centrifuge the inflammasome-stimulated plate(s) for 7 min at 300 × *g* at 4 °C.
18. Carefully transfer the uppermost 200 μL of the supernatants to a fresh 96-well plate using a multichannel pipette (*see* **Note 13**).
19. Discard the remaining supernatant in the inflammasome-stimulated plate(s) by decanting the contents into the sink and then, without flipping the plate back, tap the plate once firmly on a stack of paper towels to wick away excess supernatant without losing cells.
20. Add 200 μL of cold PBS to each well, centrifuge for 7 min at 300 × *g* at 4 °C, and repeat **step 19**.
21. Next, prepare the cell pellet for subsequent analysis by ELISA or Western blot as follows:

    To Prepare the Cell Pellet for Analysis by ELISA

    (a) Add 200 μL of medium containing 10 % FCS.
    (b) Subject the cells to three freeze–thaw cycles by transferring the plate between a freezer (−20 or −80 °C) and a 37 °C incubator. Make sure that the medium is completely frozen or thawed after each step.

(c) After the last thaw, centrifuge the plates at 300–350 × *g* to pellet debris, transfer the cell lysate supernatant to a new plate and perform a pro-IL-1β (and potentially IL-1α) ELISA in parallel to the measurement of IL-1 in the cell culture supernatants as described in Subheading 3.2 (*see* **Note 14**).

To Prepare the Cell Pellet for Analysis by Western Blot

(a) Add 40 μL of 1× SDS sample buffer to each well.

(b) Store the plate(s) at 4 °C overnight for next day Western blot analysis or at −20 or −80 °C for longer storage.

22. Transfer up to 50 μL per well of the supernatants from the inflammasome stimulation from the fresh plate (from **step 18** of Subheading 3.1) to another fresh 96-well cell culture plate and perform a LDH release assay according to manufacturer's instructions (*see* **Note 15**).
23. Store the fresh plate with the supernatants (from **step 18** of Subheading 3.1) at 4 °C overnight for selection of Western blot samples on the next day (or at −20 or −80 °C for longer storage).

3.2 ELISA

Carry out IL-1α, IL-1β, and TNF ELISAs according to the manufacturer's instructions. This section suggests time management of your ELISA experiment(s).

1. Coat the required amounts of IL-1α, IL-1β, and TNF ELISA plates with 50 μL of capture antibody solution the day before BMDC stimulation in Subheading 3.1, according to manufacturer's specifications (*see* **Note 16**).
2. During the incubation in **step 16** of Subheading 3.1, block ELISA plates and prepare a standard dilution using a high standard of 10 ng/mL and an 11-point standard curve of 1/2 dilutions with a blank.
3. Apply supernatants (from **step 18** of Subheading 3.1) and/or freeze-thaw cell lysates (from **step 21** of Subheading 3.1) to the ELISA plate (*see* **Note 14**). Most samples must be diluted for the ELISA measurement (*see* **Note 17**).
4. Following overnight incubation of the plates, develop ELISA for IL-1β, IL-1α, and TNF according to manufacturer's recommendations (*see* **Note 18**) and analyze the results.

3.3 Western Blot

1. Based on the ELISA results, choose samples from the inflammasome stimulation for further analysis by Western blot.
2. Bring the plate with the cell lysates from **step 21** of Subheading 3.1 to room temperature. Pool the triplicates into 1.5 mL tubes (*see* **Note 19**).

3. Remove undiluted supernatants from **step 18** of Subheading 3.1 from the refrigerator/freezer. Pool 30 μL of each well of a triplicate into one well of a new 96-well plate using a multichannel pipette. Add 45 μL of 3× SDS sample buffer to each pooled well, and transfer the contents of each well to a 1.5 mL tube.
4. Incubate samples from **steps 2** and **3** of this section in a 95 °C heat block for 5 min.
5. Subject the samples to SDS-PAGE using 12 % precast gels (*see* **Note 20**), or 15 % gels, if IL-1α will be analyzed (*see* **Note 21**).
6. Transfer the proteins to a nitrocellulose membrane at 100 V for 1.5 h (*see* **Note 22**).
7. Incubate the membrane in 10–15 mL Ponceau red solution for 2 min with mild shaking to confirm successful transfer.
8. Decant Ponceau red solution and wash the membrane for 2 min in distilled water with mild shaking until excess dye is removed. Ponceau red solution can be repeatedly reused.
9. Make a photocopy or scan of the blot for your documentation, and note any loading differences between samples. Figure 2 shows an example (*see* **Note 23**).
10. Block the membranes by incubating in blocking buffer for 0.5–1 h at room temperature with mild shaking (*see* **Note 24**).
11. Discard blocking buffer, add the primary antibody (0.5–1 μg/mL in blocking buffer containing 0.05 % azide), and incubate overnight at 4 °C with mild shaking (*see* **Note 25**).
12. Remove primary antibody and store at −20 °C for reuse. Wash the blots four times in wash buffer for 5 min each, at room temperature with mild shaking.
13. Dilute the appropriate HRP-conjugated secondary antibody 1:3,000–1:10,000 in blocking buffer (no azide) and add to the blots (*see* **Note 24**).
14. Incubate for 2 h at room temperature with mild shaking.
15. Wash for 2 h in wash buffer at room temperature with mild shaking, changing the buffer at least five times during the wash.
16. Tap the membrane dry using a tissue, immediately lay it over a 1 mL drop of regular or high fidelity ECL solution (on a piece of Parafilm or plastic foil) for 1–2 min, and tap dry again (*see* **Note 26**).
17. Develop the blot using standard techniques and equipment. For troubleshooting on background problems, (*see* **Note 27**).
18. Wash blots briefly in wash buffer. Repeat **steps 11–17** with different primary and secondary antibodies. Example blots are provided in Figures 2 and 4 (*see* **Note 28**).

4 Notes

1. Several protocols are available for the generation of BMDCs [22, 23]. Most of them aim at a maximum yield of immature (but differentiated) BMDCs for use in antigen presentation and T cell activation assays. As we prime the cells for inflammasome activation anyway, we are not as concerned about their preactivation status. Approximately $0.5–1 \times 10^8$ cells per mouse can be expected. Seed cells on day 0 at a density of 5×10^5–10^6 cells per mL (typically 5–10 non-tissue culture treated 10 cm dishes per mouse). On day 2, add 10 mL of medium, on day 4, 6, and 8, remove 10 mL of cell suspension from each plate, centrifuge at 300–350 × *g* for 5–7 min and discard the medium. Resuspend the cells in the same volume of fresh BMDC medium and add back to the plates. The cell density should not exceed 2×10^6 per mL. Laboratories working with bone marrow-derived myeloid cells may choose to adhere to their established protocols for generating these cells.
2. The quality of the FCS used is critical. It may be necessary to test several batches of low endotoxin FCS to find one that gives a good yield of differentiated ($CD11c^+$) immature DCs. Regular RPMI medium without GlutaMAX™ but supplemented with 2 mM glutamine can also be used for BMDC medium.
3. Typically, inhibitors will only be used on wildtype cells. Here, Ebselen, KCl, and Z-VAD-fmk are used. Ebselen is a ROS inhibitor, sometimes referred to as a glutathione mimetic. KCl is used to prevent potassium efflux form cells by increasing the extracellular levels of potassium. Z-VAD-fmk is a pan-caspase inhibitor that irreversibly binds to the active site of caspases including caspase-1. Since the activation mechanism of NLRP3 is still unclear, some debate has arisen as to the specificity of certain inhibitors of inflammasome activation. For example, increasing extracellular potassium levels to cytoplasmic concentration has significant off-target effects, presumably due to effects on membrane polarization. Furthermore, many ROS inhibitors can also affect the activity of the transcription factor NF-κB and might therefore interfere with priming [26]. If inhibitors are to be used, it is important to include controls for off-target effects or toxicity (*see* **Note 5**) and to ensure sufficient priming before adding the inhibitor. It can also be useful to determine NLRP3 levels in cell lysates, since priming can boost NLRP3 expression [27].
4. It is useful to label the 96-well plates before the experiment in order to visualize the set-up and ensure that the calculation is correct. Leave wells empty for ELISA standards if a complete

plate of the stimulation is to be transferred to one ELISA plate. Here is an example calculation for an experiment using the stimuli listed in Subheading 2:

- *For inflammasome activation—wildtype*: (3 stimuli (ATP, nigericin, alum) + 1 unstimulated = *4*) × (3 inhibitors (Ebselen, KCl, Z-VAD-fmk) + 1 uninhibited = *4*) × (replicates = *3*) × (volume per well = *200 μL*) × (pipetting loss = *1.1*) = *10.56 mL* of cells with a density of 10^6 per mL.
- *For inflammasome activation—knockout*: (3 stimuli (ATP, nigericin, alum) + 1 unstimulated = *4*) × (replicates = *3*) × (volume per well = *200 μL*) × (pipetting loss = *1.1*) = *2.64 mL* of cells with a density of 10^6 per mL.
- *For control stimulation—wildtype*: (1 stimulus (LPS) + 1 unstimulated = *2*) × (3 inhibitors (Ebselen, KCl, Z-VAD-fmk) + 1 uninhibited = *4*) × (replicates = *3*) × (volume per well = *200 μL*) × (pipetting loss = *1.1*) = *5.28 mL* of cells with a density of 10^6 per mL.
- *For control stimulation—knockout*: (1 stimulus (LPS) + 1 unstimulated = *2*) × (replicates = *3*) × (volume per well = *200μL*) × (pipetting loss = *1.1*) = *1.32 mL* of cells with a density of 10^6 per mL.

5. To control for unexpected effects of inhibitors or gene deficiency, and also for potential differences in plating density, a good option is to do a parallel experiment in which a different signaling pathway is activated. Here, LPS-induced TNF production is used for this purpose. We suggest taking an aliquot of the same cells used for inflammasome stimulation and leave them unprimed (*see* Fig. 1). These should be plated and treated in parallel to the primed cells with the inhibitor(s). Instead of treatment with inflammasome activators, stimulate the control cells with LPS for a similar duration as the primed cells are subjected to inflammasome activators (4–6 h). Measuring TNF production from this stimulation gives an indication as to whether a compound or genetic deficiency interferes with TLR4 signaling in addition to inflammasome activation. Furthermore, determining the levels of pro-IL-1α, pro-IL-1β, and pro-caspase-1 (as well as NLRP3 or ASC) in cell lysates by Western blot will give an indication as to a possible influence on priming in comparison to actual inflammasome activation.
6. The longer the DCs are in culture, the more IL-1 and caspase-1 they secrete upon inflammasome activation. This may in part be due to the higher proportion of differentiated ($CD11c^+$) cells. However, the longer the DCs are in culture, the more cells will become adherent. Classically, these cells would be deemed either mis-differentiated cells that drifted into a macrophage lineage or preactivated, mature dendritic cells.

In terms of inflammasome activation, we have found that the adherent cells in a GM-CSF culture respond in the same way as the floating cells.

7. It is particularly important that cells are resuspended in fresh medium if cell death will be determined by LDH release assay.
8. Counting and adjusting BMDCs is subject to inaccuracy. To avoid cells aggregating or sticking to the wall of the tube, it is best to count for each mouse/genotype immediately either after pooling adherent and non-adherent cells or after resuspending the cells in fresh BMDC medium. The cells should be carefully resuspended by pipetting up and down several times with a serological pipette (not by vortexing) right before taking a sample. Aliquots should not be left in a 96-well plate for later counting, since the cells might become adherent. The cells should also be counted immediately after trypan blue is added. Count the cells again after readjusting, allowing a tolerance of ±10 %. The control stimulation will help to estimate the error derived from any differences in cell density.
9. Use 1.5 mL tubes containing the required volume of BMDC medium. Leave one tube without inhibitor and add the required amount of each inhibitor stock to the other tubes and mix well. Calculation: [(number of inflammasome stimuli + 1 control stimulus + 2 unstimulated) × (3 replicates) × (50 μL) × (1.1 for pipetting loss)]. Typical dilution factors and stock concentrations are listed in Subheading 2.1, **item 15**. Note that cell culture medium contains 5 mM KCl, while cytoplasmic concentrations are around 130 mM. Therefore, concentrations tested should stay within these boundaries and 50 mM is sufficient for inflammasome inhibition. ROS inhibitors might become spontaneously oxidized over time, so stock solutions should not be stored for extended periods of more than 6 weeks.
10. The time required for an inhibitor to be effective should be evaluated for each inhibitor individually, but should generally be kept as short as possible. It might be worth performing a time course experiment (after the optimal time point for the stimulus has been previously determined, *see* **Note 12**) to find the optimal preincubation period for optimal inhibition and minimal off-target effects or toxicity.
11. Use 1.5 mL tubes containing the required volume of BMDC medium. Leave one tube without stimulus and add the required amount of each stimulus stock to the other tubes, mix well by vortexing. Note that particulate stimuli tend to aggregate, but a short sonication (with either a bath or probe sonicator) will help to dissociate them. Calculation: [(number of mice + number of inhibitors) × (3 replicates) × (50 μL) × (1.1 for pipetting loss)]. Typical stock concentrations and dilution factors are listed in **items 13** and **14** of Subheading 2.1.

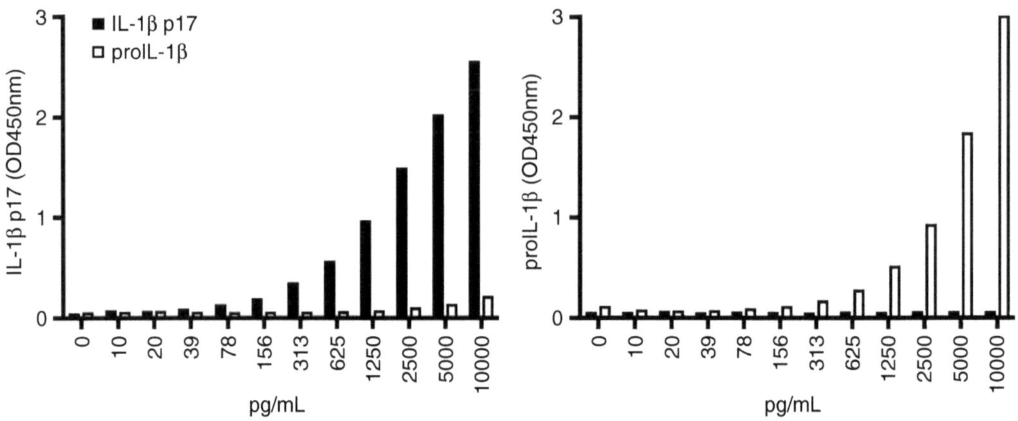

Fig. 3 Specificity of ELISAs for pro-IL-1β and IL-1β p17. Serial ½ dilutions of recombinant pro-IL-1β and IL-1β p17 were subjected to ELISA for both pro-IL-1β and IL-1β p17 according to the manufacturer's instructions. Data is presented as absorbance at 450 nm

12. The duration of the stimulation is stimulus-dependent. ATP or nigericin induce robust inflammasome activation after only 30 min to 1 h, whereas most particulate stimuli, but also many pathogens, require at least 3–4 h. It might be worth performing a time course experiment in order to identify the minimum stimulation period for full activation. Stimulation for more than 6 h should be avoided in order to minimize the effects of feedback loops, cell death, or overgrowth of a pathogen. It might be useful to harvest fast-acting stimuli earlier than slow ones, especially when measuring cell death by LDH release, since that signal will increase over time [3]. In contrast, the IL-1 signal does not change if fast-acting stimulants are left on the cells for longer time periods [3]. Therefore, in a larger experiment, samples treated with fast-acting stimuli and samples treated with slow-acting stimuli should be cultured on separate plates, while cells of different genotype or various inhibitors can be cultured on the same plate.
13. For supernatant samples that will be analyzed for IL-1 secretion, be careful not to disturb the pelleted cells. 200 μL per replicate should be enough for subsequent analysis. If the control stimulation (i.e., LPS stimulation-induced TNF production after treatment with inhibitors) was done on a separate plate, the supernatants can be diluted and the plate can be stored (at 4 °C overnight or at −20 or −80 °C for longer storage) without separating the supernatants from the cells.
14. In contrast to IL-1α, the commercially available IL-1β ELISA kits do not detect pro-IL-1β, and should therefore not be used to monitor intracellular pro-IL-1β levels [3, 28]. This is illustrated in Fig. 3. However, intracellular amounts of pro-IL-1β

can be detected by an ELISA kit designed to detect only the pro-form (eBioscience). By subjecting the cells to repeated freeze-thaw cycles, the intracellular pro-IL-1 is released and is thereby made available for ELISA testing. This allows direct comparison of the amount of pro-IL-1 inside the cells to the amount of secreted IL-1 in the supernatants.

15. A triplet of fresh BMDC medium should be included to determine the background of the medium without cells. Freeze-thaw or detergent-treated cells can be used to determine 100 % lysis (maximal LDH release).

16. One well in the stimulations corresponds to one well on the ELISA plate. Therefore, if wells are left empty on the stimulation plates for ELISA standards, the same number of plates is needed for the TNF and IL-1 ELISAs as are needed for the control stimulation and the inflammasome stimulation, respectively. The final triplets measured are replicates of the whole experimental procedure, not only of the ELISA method.

17. Many stimuli may induce the production of more than 10 ng/mL of IL-1 or TNF. This is above the measurable range of the ELISA, so the supernatants should be diluted at least 1/3 for measurement. Greater dilutions (1/10) may be necessary for some experiments. For cytokine measurement from the cell pellet, the samples should be diluted at 1/10.

18. The last washing steps after incubation with streptavidin-HRP and before adding the substrate are critical in reducing the background and minimizing variation between replicate samples.

19. The cell lysates serve two purposes: They represent a control to demonstrate that equal amounts of cells (e.g., from different genotypes) were used. Some stimuli like nigericin or ATP are quite toxic to the cells, which might reduce the total protein content in some samples. In addition, the pool of intracellular pro-IL-1 and pro-caspase-1 and potential intracellular cleavage can be monitored. Pool the lysates from each triplet in one 1.5 mL tube. The samples are viscous until they have been boiled.

20. We use the Mini-PROTEAN gel system from Bio-Rad with 1 mm thick 12 % precast gels and 15 lanes per gel and the related wet blot system and run the gels at 100–160 V for 1–2 h until just before the dye front leaves the gel. We have also obtained good results with precast gradient gels from Invitrogen.

21. The predicted protein size of both mature IL-1β and cleaved IL-1α are 17.4 kDa and 18.0 kDa, respectively. However, the molecular weight of cleaved IL-1α has variably been reported to be between 15 and 17 kDa, while mature IL-1β is usually labelled as 17 kDa. Figure 4 shows a direct comparison between

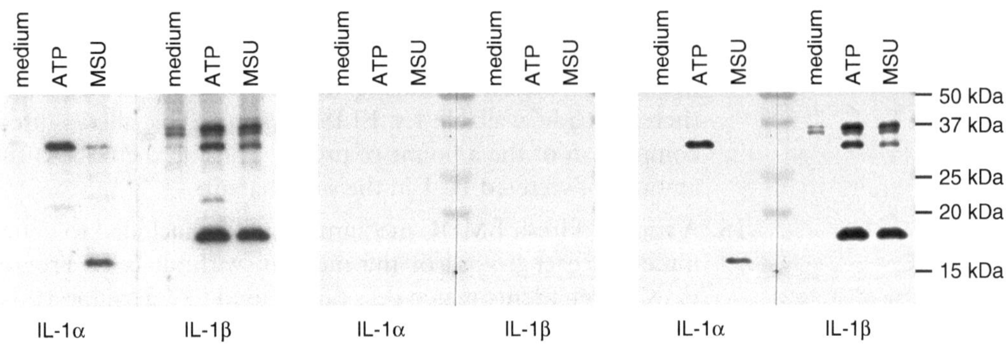

Fig. 4 Comparison between the migration of murine IL-1α and IL-1β in an SDS-PAGE. Murine BMDCs were left unstimulated or stimulated with ATP or monosodium urate crystals (MSU). The supernatant samples were loaded twice on the same 15 % gel and blotted on a nitrocellulose membrane. This was cut in two after transfer; one half was probed for IL-1α, the other for IL-1β. For the exposure to film, the two halves of the blot were arranged side by side so that the molecular weight markers lined up. The blots themselves were scanned directly after exposure to the film and in the exact same arrangement. A scan of the developed films (*left*) and a scan of the membranes (*middle*) were overlaid (*right*). The IL-1α antibody has been raised against the full C-terminus of murine IL-1α (aa 99–268) [30]

the migration of murine IL-1α and IL-1β in an SDS-PAGE and Western blot from cell culture supernatants. Murine IL-1α migrates below its predicted molecular weight (at ~16 kDa), while IL-1β migrates at approximately the same level as predicted. This could be because IL-1α has a low theoretical isoelectric point (5.57) and is thus more negatively charged than IL-1β (pI 8.08) in the pH range that is commonly used in the separating gel of an SDS-PAGE. For such a small protein, this additional negative charge (on top of that provided by the SDS) could be sufficient to shift the charge to mass ratio of the unfolded peptide and allow IL-1α to migrate faster in the gel. Similarly, human IL-1α and IL-1β have different isoelectric points, and it was this difference that allowed the initial identification of the two IL-1 cytokines as distinct proteins [29]. The different behavior in an SDS-PAGE has consequences when measuring IL-1α along with IL-1β and caspase-1 in a Western blot. We suggest here to use precast 12 % SDS PAGE gels for the analysis of IL-1β and caspase-1. These gels are easy to use and allow maximal resolution of the pro- and mature forms of these two proteins. However, IL-1α will migrate in the running front of a 12 % gel, so 15 % gels should instead be used. These can be purchased (e.g., from Life Technologies) or made as previously described [22].

22. A wet blot system may lead to better transfer than a semidry system, and thus a stronger signal. The current should be no higher than 300 mA per tank. The blotting buffer might get

warm, but should not get hot. The cold packs that come with the system should be used to keep the temperature moderate. Alternatively, 2–3 water-filled 50 mL tubes frozen at −20 °C can be used as cold packs. Make sure that the blot buffer is thoroughly mixed before filling the tanks, since poor mixing might increase the resistance. If Western blots are performed infrequently, the blot buffer should be made without ethanol and it should be freshly added before each blot to avoid evaporation of the ethanol.

23. The most prominent band in the supernatants is around 60 kDa, and corresponds to the albumin in the FCS.
24. The volumes of blocking, washing and antibody solutions will vary greatly depending on the number of blots, the size of the blots, and the size of the vessel in which the incubations are performed. Washing and blocking steps should be carried out using generous volumes. Antibody incubation steps should use no more volume than is necessary to cover the blots and allow them to move freely against one another.
25. Primary antibody solutions in blocking buffer with azide can be used until they are exhausted (i.e., when the signal becomes weaker and requires the use of stronger ECL solutions for detection). Antibody solutions should be stored at −20 °C in the meantime.
26. Do not let the blots air-dry completely after tapping them but quickly cover them in transparent foil and move them to a developing cassette or put them back into ECL or wash buffer. Use tissues or paper towels that have no pattern pressed into them, as this might lead to uneven removal of wash buffer or ECL solution. If normal ECL is too weak, but high sensitivity ("femto") is too strong, femto, 1:3 diluted in normal ECL solution or water can be used. If a CCD camera system is used for documentation, the last tapping-dry step can be omitted and instead, the ECL-soaked membrane can be transferred to the CCD system without any covering foil to avoid potential scattering of the light signal by the foil.
27. Several methods for minimizing background are described here. First, longer or more intense washing steps could be performed. We have found washing overnight at 4 °C to be helpful in some cases. Using a different wash buffer, containing 0.1 % Triton X-100 and 1 % skim milk powder in TBS or PBS seems to help for some antibodies, especially if used for the whole process from the first blocking of the membrane up to the second-last wash step before adding ECL solution (the last wash before ECL should be done without milk in the buffer). The use of a lower concentration of primary and/or secondary antibodies, or a different secondary antibody, directed against the specific isotype of the primary antibody, may also help.

28. As the common antibodies used (Figures 2 and 4) were generated in different species (caspase-1: mouse, IL-1β: goat, IL-1α: hamster), the same membrane can be probed for all three proteins. The order in which these three antibodies are used should be decided based on signal strength and the project-specific interest. Either the weakest signal or the most important readout for a given experimenter should be measured first.

Acknowledgments

OG is supported by the Bavarian Ministry of Sciences, Research and the Arts in the Framework of the Bavarian Molecular Biosystems Research Network. C.J.T. is supported by an NSERC postgraduate scholarship. K.S.S. is supported by a Ph.D. scholarship by the faculty of medicine.

References

1. Martinon F, Petrilli V, Mayor A, Tardivel A, Tschopp J (2006) Gout-associated uric acid crystals activate the NALP3 inflammasome. Nature 440(7081):237–241. doi:nature04516 [pii], 10.1038/nature04516
2. Fettelschoss A, Kistowska M, LeibundGut-Landmann S, Beer HD, Johansen P, Senti G, Contassot E, Bachmann MF, French LE, Oxenius A, Kundig TM (2011) Inflammasome activation and IL-1beta target IL-1alpha for secretion as opposed to surface expression. Proc Natl Acad Sci USA 108(44):18055–18060. doi:10.1073/pnas.1109176108
3. Gross O, Yazdi AS, Thomas CJ, Masin M, Heinz LX, Guarda G, Quadroni M, Drexler SK, Tschopp J (2012) Inflammasome activators induce interleukin-1alpha secretion via distinct pathways with differential requirement for the protease function of caspase-1. Immunity 36(3):388–400. doi:10.1016/j.immuni.2012.01.018
4. Keller M, Ruegg A, Werner S, Beer HD (2008) Active caspase-1 is a regulator of unconventional protein secretion. Cell 132(5):818–831. doi:10.1016/j.cell.2007.12.040
5. Kuida K, Lippke JA, Ku G, Harding MW, Livingston DJ, Su MS, Flavell RA (1995) Altered cytokine export and apoptosis in mice deficient in interleukin-1 beta converting enzyme. Science 267(5206):2000–2003
6. Dinarello CA (2009) Immunological and inflammatory functions of the interleukin-1 family. Annu Rev Immunol 27:519–550. doi:10.1146/annurev.immunol.021908.132612
7. Sims JE, Smith DE (2010) The IL-1 family: regulators of immunity. Nat Rev Immunol 10(2):89–102. doi:10.1038/nri2691
8. Nickel W, Seedorf M (2008) Unconventional mechanisms of protein transport to the cell surface of eukaryotic cells. Annu Rev Cell Dev Biol 24:287–308. doi:10.1146/annurev.cellbio.24.110707.175320
9. Carruth LM, Demczuk S, Mizel SB (1991) Involvement of a calpain-like protease in the processing of the murine interleukin 1 alpha precursor. J Biol Chem 266(19):12162–12167
10. Kobayashi Y, Yamamoto K, Saido T, Kawasaki H, Oppenheim JJ, Matsushima K (1990) Identification of calcium-activated neutral protease as a processing enzyme of human interleukin 1 alpha. Proc Natl Acad Sci USA 87(14):5548–5552
11. Schroder K, Tschopp J (2010) The inflammasomes. Cell 140(6):821–832. doi:S0092-8674(10)00075-9 [pii], 10.1016/j.cell.2010.01.040
12. Martinon F, Burns K, Tschopp J (2002) The inflammasome: a molecular platform triggering activation of inflammatory caspases and processing of proIL-beta. Mol Cell 10(2):417–426. doi:S1097276502005993 [pii]
13. Dostert C, Petrilli V, Van Bruggen R, Steele C, Mossman BT, Tschopp J (2008) Innate immune activation through Nalp3 inflammasome sensing of asbestos and silica. Science 320(5876):674–677. doi:1156995 [pii], 10.1126/science.1156995
14. Henderson C, Goldbach-Mansky R (2010) Monogenic autoinflammatory diseases: new

insights into clinical aspects and pathogenesis. Curr Opin Rheumatol 22(5):567–578. doi:10.1097/BOR.0b013e32833ceff4, 00002281-201009000-00017 [pii]
15. Ghiringhelli F, Apetoh L, Tesniere A, Aymeric L, Ma Y, Ortiz C, Vermaelen K, Panaretakis T, Mignot G, Ullrich E, Perfettini JL, Schlemmer F, Tasdemir E, Uhl M, Genin P, Civas A, Ryffel B, Kanellopoulos J, Tschopp J, Andre F, Lidereau R, McLaughlin NM, Haynes NM, Smyth MJ, Kroemer G, Zitvogel L (2009) Activation of the NLRP3 inflammasome in dendritic cells induces IL-1beta-dependent adaptive immunity against tumors. Nat Med 15(10):1170–1178. doi:nm.2028 [pii], 10.1038/nm.2028
16. Masters SL, Dunne A, Subramanian SL, Hull RL, Tannahill GM, Sharp FA, Becker C, Franchi L, Yoshihara E, Chen Z, Mullooly N, Mielke LA, Harris J, Coll RC, Mills KH, Mok KH, Newsholme P, Nunez G, Yodoi J, Kahn SE, Lavelle EC, O'Neill LA (2010) Activation of the NLRP3 inflammasome by islet amyloid polypeptide provides a mechanism for enhanced IL-1beta in type 2 diabetes. Nat Immunol 11(10):897–904. doi:ni.1935 [pii], 10.1038/ni.1935
17. Duewell P, Kono H, Rayner KJ, Sirois CM, Vladimer G, Bauernfeind FG, Abela GS, Franchi L, Nunez G, Schnurr M, Espevik T, Lien E, Fitzgerald KA, Rock KL, Moore KJ, Wright SD, Hornung V, Latz E (2010) NLRP3 inflammasomes are required for atherogenesis and activated by cholesterol crystals. Nature 464(7293):1357–1361. doi:nature08938 [pii], 10.1038/nature08938
18. Hornung V, Ablasser A, Charrel-Dennis M, Bauernfeind F, Horvath G, Caffrey DR, Latz E, Fitzgerald KA (2009) AIM2 recognizes cytosolic dsDNA and forms a caspase-1-activating inflammasome with ASC. Nature 458(7237):514–518. doi:nature07725 [pii], 10.1038/nature07725
19. Poeck H, Bscheider M, Gross O, Finger K, Roth S, Rebsamen M, Hannesschlager N, Schlee M, Rothenfusser S, Barchet W, Kato H, Akira S, Inoue S, Endres S, Peschel C, Hartmann G, Hornung V, Ruland J (2010) Recognition of RNA virus by RIG-I results in activation of CARD9 and inflammasome signaling for interleukin 1 beta production. Nat Immunol 11(1):63–69. doi:ni.1824 [pii], 10.1038/ni.1824
20. Boyden ED, Dietrich WF (2006) Nalp1b controls mouse macrophage susceptibility to anthrax lethal toxin. Nat Genet 38(2):240–244. doi:ng1724 [pii], 10.1038/ng1724
21. Mariathasan S, Newton K, Monack DM, Vucic D, French DM, Lee WP, Roose-Girma M, Erickson S, Dixit VM (2004) Differential activation of the inflammasome by caspase-1 adaptors ASC and Ipaf. Nature 430(6996):213–218. doi:10.1038/nature02664, nature02664 [pii]
22. Gross O (2012) Measuring the inflammasome. Methods Mol Biol 844:199–222. doi:0.1007/978-1-61779-527-5_15
23. Lutz MB (2004) IL-3 in dendritic cell development and function: a comparison with GM-CSF and IL-4. Immunobiology 209(1–2):79–87
24. Manzanero S (2012) Generation of mouse bone marrow-derived macrophages. Methods Mol Biol 844:177–181. doi:10.1007/978-1-61779-527-5_12
25. Palleroni AV, Varesio L, Wright RB, Brunda MJ (1991) Tumoricidal alveolar macrophage and tumor infiltrating macrophage cell lines. Int J Cancer 49(2):296–302
26. Murphy MP, Holmgren A, Larsson NG, Halliwell B, Chang CJ, Kalyanaraman B, Rhee SG, Thornalley PJ, Partridge L, Gems D, Nystrom T, Belousov V, Schumacker PT, Winterbourn CC (2011) Unraveling the biological roles of reactive oxygen species. Cell Metab 13(4):361–366. doi:10.1016/j.cmet.2011.03.010
27. Bauernfeind FG, Horvath G, Stutz A, Alnemri ES, MacDonald K, Speert D, Fernandes-Alnemri T, Wu J, Monks BG, Fitzgerald KA, Hornung V, Latz E (2009) Cutting edge: NF-kappaB activating pattern recognition and cytokine receptors license NLRP3 inflammasome activation by regulating NLRP3 expression. J Immunol 183(2):787–791. doi:jimmunol.0901363 [pii], 10.4049/jimmunol.0901363
28. Wewers MD, Pope HA, Miller DK (1993) Processing proIL-1 beta decreases detection by a proIL-1 beta specific ELISA but increases detection by a conventional ELISA. J Immunol Methods 165(2):269–278
29. Dinarello CA, Goldin NP, Wolff SM (1974) Demonstration and characterization of two distinct human leukocytic pyrogens. J Exp Med 139(6):1369–1381
30. Fuhlbrigge RC, Sheehan KC, Schreiber RD, Chaplin DD, Unanue ER (1988) Monoclonal antibodies to murine IL-1 alpha. Production, characterization, and inhibition of membrane-associated IL-1 activity. J Immunol 141(8):2643–2650

Chapter 11

Reconstituting the NLRP1 Inflammasome In Vitro

Benjamin Faustin and John C. Reed

Abstract

Nucleotide-binding domain leucine-rich-repeat containing receptors; NOD-Like Receptors (NLRs) were originally described as microbial sensors involved in host defense against pathogens that comprise an important component of the innate immune system. Recently, their cellular functions have expanded beyond classical pathogen detection, to danger sensors that may contribute to the pathophysiology of a wide range of inflammation-driven human illnesses such as metabolic diseases (atherosclerosis, obesity, type 2 diabetes, gout, age-related macular degeneration) and neurological disorders (Alzheimer's disease). Pathogen-stimulated NLRs such as NLR family Pyrin domain-containing protein 1 (NLRP1) assemble into molecular platforms called "inflammasomes" to activate inflammatory protease caspase-1, which processes pro-IL-1β and pro-IL-18 into active cytokines. We describe methods for reconstituting the human NLRP1 inflammasome in vitro. Protocols are provided for: (a) expression and purification of inflammasome core components (NLRP1 and pro-caspase-1 proteins) using the baculovirus/insect cell expression system, and (b) functional monitoring of NLRP1-mediated caspase-1 activation in response to NLRP1 ligand muramyl dipeptide (MDP) and ATP.

Key words Inflammasome, NLRP1, Caspase-1, Recombinant proteins

1 Introduction

Inflammasomes are multi-protein complexes that stimulate caspase-1 activation. These protein complexes initiate inflammatory signaling in response to molecular determinants produced by pathogens and tissues injury [1, 2]. Two types of inflammasomes have been described, involving either nucleotide-binding domain leucine-rich-repeat containing receptors; NOD-like receptor (NLRs) family proteins or the PYRIN family protein, absent in melanoma 2 (AIM2)-like receptors as anchoring scaffolds onto which protein complex assembly occurs. While playing critical roles in host defense against pathogens, functions of NLRs have also been associated with many diseases, including atherosclerosis, gout, obesity, type 2 diabetes, age-related macular degeneration, and Alzheimer's disease [3–7]. Most NLR family members contain a nucleotide-binding NACHT domain and varying numbers of

Christine M. De Nardo and Eicke Latz (eds.), *The Inflammasome: Methods and Protocols*, Methods in Molecular Biology, vol. 1040, DOI 10.1007/978-1-62703-523-1_11, © Springer Science+Business Media New York 2013

leucine-rich repeat (LRR) domains, often combined with a variety of protein interaction domains [1, 2]. Interaction of microbial and other types of ligands with the LRRs allows the NACHT domain-mediated oligomerization of NLRs, thus creating a platform for recruitment and activation of downstream effectors. In particular, some NLRs can interact directly with pro-caspase-1 through caspase activation and recruitment domain (CARD) interactions, or indirectly via the adapter protein apoptosis-associated speck-like protein containing a card domain (ASC), which also contains a Pyrin domain (PYD) [1]. Caspase-1 belongs to the inflammatory group of caspases, which cleave pro-interleukin-1β (IL-1β) and pro-IL-18 [8] to produce active cytokines that prime host adaptive immunity and stimulate inflammatory responses. Excessive activation of caspase-1 can also induce immune cell death, either by apoptosis [9] or by a variant termed "pyroptosis," especially during host responses to pathogens [4].

NLR family, Pyrin domain-containing protein 1(NLRP1) is one of 22 NLRs expressed in humans [10]. Hereditary polymorphisms within the NLRP1 gene have been associated with Vitiligo and related autoimmune diseases [11]. This NLR member undergoes auto-proteolytic processing when expressed in mammalian cells, through a highly conserved ZU5-Like domain (located between LRR and CARD domains), and previously described as the "Function to Find Domain" (FIIND) [12, 13]. This mechanism may regulate immune responses and have critical implications in NLRP1-associated autoimmune diseases. NLRP1 was the first inflammasome to be reconstituted in vitro [14]. This work has helped to define the sequence of biochemical events required for inflammasome assembly and subsequent caspase-1 activation, revealing a two-step activation mechanism [14]. The in vitro reconstituted NLRP1 inflammasome system has also provided insights into the molecular mechanism by which anti-apoptotic proteins Bcl-2 and Bcl-XL inhibit NLRP1 [15, 16]. Furthermore, it has aided in the understanding of the inhibitory activity of certain viral antagonists of NLRs, such as the ORF63 protein of Kaposi Sarcoma Virus (Human Herpes Virus 8, HHV8).

Herein, we describe the detailed methodology for NLRP1 inflammasome reconstitution in vitro, by purifying NLPR1 and pro-caspase-1 core components (using the baculovirus-insect cell expression system), and detail how to combine these recombinant proteins to monitor NLRP1-mediated caspase-1 activation and ATP binding. In addition to NLRP1, this procedure may serve as a prototypical method to discover biochemical regulators of other inflammasomes, including chemical agonists and antagonists.

2 Materials

Prepare all solutions using ultrapure water and analytical grade reagents. Prepare and store all reagents at room temperature (unless indicated otherwise). Diligently follow all waste disposal regulations when disposing waste materials. Insect cell culture medium should always be supplemented with l-Glutamine and Penicillin/Streptomycin.

2.1 Generating Recombinant Baculoviruses by Co-transfection

1. BD Biosciences BaculoGold protein expression system containing BaculoGold™ Linearized Baculovirus DNA.
2. Transfection Buffer A and B set (BD).
3. Human NLRP1 and pro-caspase-1 genes subcloned into pAc-GHLT or pAcHLT baculovirus transfer vectors of the BD BaculoGold protein expression system (empty transfer vectors are purchased from BD Biosciences) (*see* **Note 1**).
4. Low passage Sf9 cells (frozen batch from BD).
5. Sf900III media (Invitrogen).
6. 10 and 15 cm tissue culture plates.
7. Sterile 10 mL pipettes.
8. Sterile 1.5 mL tubes.
9. Sterile 15 mL tubes.
10. Hemocytometer.
11. Inverted light microscope.
12. 27 °C incubator.
13. 6-well tissue culture plates.
14. Centrifuge with swinging buckets set at room temperature, holding 15 mL tubes, convenient for 3,000 ×*g* spinning rate.

2.2 Protein Expression

1. Low passage Sf9 cells.
2. Sf900III media (Invitrogen).
3. Recombinant baculoviruses.
4. 27 °C incubator.
5. Hemocytometer.
6. Spinner flask.
7. Spinner apparatus.
8. SDS-PAGE sample buffer (Sigma).

2.3 Protein Purification

1. 300 mL Nalgene bottles.
2. Stir bar.
3. High-speed centrifuge set at 4 °C, holding 300 mL Nalgene bottles and convenient for 3,000 and 25,000 ×*g* spinning rates.

4. Insect cell lysis buffer A for GST-His6-NLRP1: 10 mM, Tris–HCl, pH 7.5, 130 mM NaCl, 1 % Triton-X-100, 10 mM NaF, 10 mM NaPi, 10 mM NaPPi. Store at 4 °C.
5. Insect cell lysis buffer B for His6-NLRP1 or His6-pro-caspase-1: 20 mM Hepes-KOH, pH 7.5, 10 mM KCl, 1 mM EGTA, 1 mM EDTA, 1.5 mM $MgCl_2$, 1 mM DTT, 130 mM NaCl, 1 % Triton X-100. Store at 4 °C.
6. Mammalian protease inhibitors (Sigma) supplemented with 50 mM PMSF.
7. 50 mL tubes.
8. Centrifuge with swinging buckets set at 4 °C, holding 50 mL tubes, convenient for 6,000 × g spinning rate.
9. Low-speed rotator at 4 °C.
10. High Binding capacity Ni^{2+}-NTA agarose or Glutathione-sepharose 4B (Amersham).
11. His-wash buffer: 20 mM Hepes-KOH, pH 7.5, 10 mM KCl, 1 mM EGTA, 1 mM EDTA, 1.5 mM $MgCl_2$, 1 mM DTT, 1 M NaCl, 20 mM imidazole.
12. Nickel chromatography elution buffer: Insect cell lysis buffer B + 250 mM imidazole.
13. GST chromatography elution buffer: 50 mM Tris–HCl, pH 8.0, 10 mM GSH, 1 % Triton X-100.
14. AKTA Purifier set up with Superdex-200 24 mL column (Amersham).
15. Fraction collector set with 400 μL volume fraction.
16. Gel Filtration Buffer C: 20 mM Hepes-KOH, pH 7.5, 10 mM KCl, 1 mM EGTA, 1.5 mM $MgCl_2$, 1 mM DTT, 150 mM NaCl, 0.1 % CHAPS.
17. Gel filtration protein standards (Bio-Rad, range of separation from 670 to 1.3 kDa_{app} molecular weights).
18. Millipore Eppendorf concentrators, 50 kDa and 10 kDa cut-offs for NLRP1 and pro-caspase-1 respectively.
19. SDS-PAGE and Western blot apparatus.
20. UV laser scanner for gel imaging.
21. SYPRO Ruby protein gel stain.
22. Monoclonal anti-GST antibody (Sigma).
23. Monoclonal anti-caspase-1 p45 antibody (Imgenex).

2.4 In Vitro NLRP1 Inflammasome Assays

1. Black, flat-bottom 96-well plate, 20 μL total volume.
2. Sterile double distilled water (DDW).
3. 2× caspase-1 activity buffer: 100 mM HEPES, pH 7.5, 10 % sucrose, 0.1 % CHAPS, 1 mM dithiothreitol (DTT) and the same buffer but with 10 mM DTT.

4. Bovine serum albumin (BSA).
5. High Throughput microplate reader (Analyst AD Assay Detection System LJL Biosystems) with standard settings to perform fluorescence-based assays in kinetic mode and with a high level of sensitivity. In addition, it should be able to perform Fluorescent Polarization Assay.
6. NLR ligands: Muramyl dipeptide MDP-LD (active enantiomer, from Alexis Biochemicals), MDP-DD (inactive enantiomer, from Alexis Biochemicals), l-alanyl-γ-D-glutamyl-meso-diaminopimelic acid: γ -tri-DAP, 1 μM final (Alexis Biochemicals).
7. 0.5 mM ATP and $MgCl_2$.
8. Caspase-1 fluorescent substrate Ac-WEHD-AMC (Calbiochem).
9. Caspase-1 inhibitor Ac-YVAD-fmk (Calbiochem).
10. Fluorescein-conjugated ATP analog (Fluorescein-12-ATP, 10 nM final, from PerkinElmer) (*see* Fig. 3 for detailed structure).

3 Methods

Carry out all procedures at room temperature unless otherwise specified.

3.1 Expression of Inflammasome Proteins Using the Baculovirus-Insect Cell Expression System

3.1.1 Recombinant Virus Preparation

1. Seed 1.5×10^6 Sf9 cells/well (~60 % confluency) in a 6-well plate (*see* **Note 2**). Prepare another plate in the same way; this will serve as the negative control (non-transfected).
2. Allow 1 h for the cells to adhere firmly to the plate.
3. Follow BD Biosciences protocol for the transfection procedure with minor modifications as below.
4. Remove the culture medium from the plate and add 1 mL of Transfection Buffer A (*see* **Note 3**).
5. For the generation of recombinant baculoviruses carrying pro-caspase-1, mix 0.5 μg of BaculoGold™ Linearized Baculovirus DNA with 5 μg recombinant baculovirus transfer vector DNA containing pro-caspase-1 per co-transfection in a 1.5 mL tube. For the generation of recombinant baculoviruses carrying NLRP1, mix 1 μg of BaculoGold™ Linearized Baculovirus DNA with 10 μg recombinant baculovirus transfer vector DNA containing NLRP1 per co-transfection in a 1.5 mL tube (*see* **Note 4**).
6. Mix each DNA solution with 1 mL of the Transfection Buffer B.
7. Add 1 mL of the Transfection Buffer B/DNA solution dropwise to the insect cells on the tissue culture plate. Gently rock

the plate back and forth after every 2–3 drops to mix the newly added solution with the Transfection Buffer A. During this procedure, a fine milky precipitate should form.

8. To express GST-6His-NLRP1 or 6His-pro-caspase-1, transfect transfer vectors pAcGHLT-NLRP1 or pAcHLT-pro-caspase-1 mixed with BaculoGold™ Linearized DNA.
9. Incubate the plates at 27 °C for 4 h.
10. After 4 h, remove the medium from the wells. Add 2 mL of fresh Sf900III insect medium per well and incubate the cells at 27 °C for 4–5 days.
11. After 4–5 days, look at the cells with an inverted microscope and check for signs of infection (*see* **Note 5**).
12. Collect the cell culture by tapping on the plate, cells should detach easily.
13. Centrifuge the culture in 15 mL tubes for 5 min at 3,000 × *g*, and keep the 2 mL supernatant. This is SUP#1 with a low virus titer (keep at 4 °C at dark).

3.1.2 Virus Amplification

After co-transfection, the recombinant baculovirus must be amplified to obtain a high titer stock solution.

1. Seed 5×10^6 Sf9 cells (~60 % confluency) in a 10 cm plate.
2. Allow 1 h for the cells to attach.
3. Add the 2 mL SUP#1 (transfection supernatant from **step 13** of Subheading 3.1.1) to the plate. At this stage, Multiplicity Of Infection (MOI) should be <1.
4. Incubate the cells at 27 °C for 4–5 days to allow viral particles to be released into the medium.
5. After 2–3 days, inspect the cells for signs of infection (*see* **Note 6**).
6. After 4–5 days, collect the cell culture and centrifuge in 15 mL tubes for 5 min at 3,000 × *g*.
7. Collect the 12 mL supernatant. This is SUP#2 with a high virus titer of ~2×10^8 pfu/mL. Keep 1 mL aliquots at 4 °C (at dark) for 1 month if needed for small scale protein expression, and freeze the rest at −80 °C for longer periods of time (stable for several months).

3.1.3 Small-Scale Protein Expression

This experiment should only be performed once, to make sure that recombinant proteins are well expressed. A rough purification is performed to get an estimate of protein recovery and yield.

1. Seed 10^7 Sf9 cells in a 15 cm plate.
2. Allow 1 h for the cells to adhere.

3. Add various amounts of SUP#2 from **step 7** of Subheading 3.1.2 (from 200 μL to 2 mL, typically ranging 1 < MOI < 10).
4. Incubate cells at 27 °C for 48 h to allow for infection.
5. Centrifuge the cells for 5 min at 3,000 × *g*, discard the supernatant, then lyse pellets with 500 μL lysis Buffer A (GST) or B (His) and perform a pre-purification step by batch using epitope tags and Glutathione or Ni^{2+}-NTA beads (50 μL) for 2 h at 4 °C, respectively.
6. Centrifuge the beads for 10 min at 6,000 × *g*, discard the supernatant, and add 50 μL SDS-PAGE sample buffer, and boil for 5 min at 95 °C.
7. Centrifuge the beads for 10 min at 6,000 × *g*, load the supernatant on SDS-PAGE and verify protein expression by Western blot using anti-GST and anti-pro-caspase-1 antibodies.
8. For larger expression batches, infect three/four 15 cm plates at 2×10^7 cells/plate with 10 μL SUP#2 high titer virus (typically MOI < 1). Incubate for 4–5 days, and inspect after 2–3 days to detect signs of infection (*see* **Note 6**).
9. Centrifuge the culture for 5 min at 3,000 × *g* and collect the supernatant. This is SUP#3. Infect 2 L of fresh Sf9 cells cultured in suspension flasks with SUP#3 at the desired MOI determined during expression trials (*see* Subheading 3.1.3 and **Note 7**). At 48 h post-infection, spin the culture down and keep cell pellets for protein purification.

3.2 Purification of Inflammasome Proteins

A two-step purification is performed to purify the inflammasome components NLRP1 and pro-caspase-1. In the first step, the proteins are purified by affinity purification according to their epitope tag. GST-tagged proteins are purified using glutathion chromatography and 6His-tagged proteins are enriched using nickel affinity chromatography. The second purification step is a size exclusion chromatography (*see* **Note 8**).

1. Lyse cells in 150 mL per 2 L of infected Sf9 cells with insect cell lysis Buffer A or B (depending on whether a GST-tagged or 6His-tagged protein is purified) for 1 h at 4 °C under gentle agitation with a stir bar.
2. Centrifuge the cells at 25,000 × *g* for 20 min at 4 °C to clear the cell lysate from nuclei.
3. Collect the post-nuclear supernatants and divide them into three 50 mL tubes for batch purification (50 mL/tube).
4. Incubate each tube with 500 μL of glutathione-Sepharose 4B for GST-tagged proteins or 1 mL of Ni^{2+}-NTA beads for 6His-tagged proteins, at 4 °C for 2 h on a low-speed rotator.

5. Wash and elute GST-tagged proteins as follows: Collect GSH-sepharose by centrifugation at 6,000 × *g* for 10 min at 4 °C. Wash the resin 6× by incubating the beads for 10 min with 20 mL lysis Buffer A on a rotator and then centrifuge at 6,000 × *g* for 10 min at 4 °C.
6. After the final wash, remove the wash buffer and elute the GST-His6-NLRP1 protein by incubating the beads for 10 min in 5 mL of GST chromatography elution buffer at room temperature. Repeat this step a second time.
7. Collect the supernatant containing the eluted proteins and concentrate to a volume of 1 mL using a 50 kDa cut-off filtration device by centrifugation at 6,000 × *g* for as long as required to reach a 1 mL volume in the filtrate.
8. Wash and elute His-tagged proteins as follows: Transfer Ni^{2+}-NTA beads into a micro-column (10 mL capacity) connected to a pump and wash beads overnight at low pump speed using at least 400 mL His wash buffer per 1.5 mL of resin. Then, remove excess salt by washing with 100 mL lysis buffer B.
9. After the wash, elute proteins with nickel chromatography elution buffer (same volume as the packed beads) over a time of 10 min (no DTT added in buffer for pro-caspase-1) (*see* **Note 9**). Recover the beads by centrifugation for 10 min, 6,000 × *g* at 4 °C. Concentrate His6-NLRP1 with a 10-kDa cut-off filtration device to 1 mL.
10. Size exclusion chromatography: Equilibrate a Superdex 200 column using at least two column volumes of Buffer C. Then, apply gel filtration protein standards and run the instrument with a flow rate of 0.5 mL/min, and record the elution profile (*see* **Note 10**). Obtain a standard curve by plotting absorbance values (at 280 nm) for each protein peak as a function of the corresponding volume of elution. Keep the same instrument settings and then perform size exclusion chromatography with the NLRP1 or pro-caspase-1 protein containing samples.
11. For the GST-6His-NLRP1 purification, inject 1 mL of the protein-containing sample and perform the size exclusion chromatography with Buffer C, with a flow rate of 0.5 mL/min and collect 400 μL fraction volumes. Monitor the eluted proteins with the absorbance set at 280 nm. After all proteins have been eluted from the column (absorbance values constant and back to background), collect the fractions corresponding to ~150 kDa_{app} monomers (spot the peak corresponding to the elution volume necessary to elute ~150 kDa_{app} proteins) (*see* **Note 11**).
12. Pool fractions and concentrate up to 200 μL using 50 kDa cut-off filtration devices. Determine the protein concentration using a Bradford assay, and then quickly freeze aliquots in dry ice/ethanol and store at −80 °C.

13. Perform His6-NLRP1 or His6-pro-caspase-1 size exclusion chromatography using Buffer C as described in **step 10** and **11** of Subheading 3.2 (avoid DTT in Buffer C for pro-caspase-1) (*see* **Note 9**). Collect ~50 kDa_{app} molecular weight fractions containing the monomeric pro-caspase-1 protein (*see* **Note 12**). Determine the protein concentration using a Bradford assay and then adjust to 50 µg/mL, aliquot samples, and quickly freeze in dry ice/ethanol, and store at −80 °C.
14. Analyze each purification batch for protein size and purity by SDS-PAGE and protein staining (*see* **Note 13**).

3.3 In Vitro Inflammasome Functional Assays

3.3.1 Caspase-1 Activation Measurements

First, perform a "calibration" experiment with pro-caspase-1 monomers (which must be performed for each new batch of purified protein). The aim is to obtain the amount of pro-caspase-1 that will not be activated by autoproteolysis, but leads to an NLRP1-dependent five- to sixfold increase of caspase-1 activity in the presence of activators MDP/ATP/Mg^{2+}, using the inflammasome reconstitution assay (*see* **Note 14**).

The assay reactions are performed in black flat-bottom 96-well plates (20 µL total volume) as follows:

1. From the pro-caspase-1 purified monomers, prepare twofold serial dilutions from a 50 µg/mL stock solution, in Buffer C and add 2 µL/well. Maintain total protein constant by adding increasing amounts of BSA to decreasing amounts of pro-caspase-1. Prepare sufficient amounts to test two triplicates for each concentration.
2. Add 2 µL/well of 2× caspase-1 activity buffer with 1 or 10 mM DTT. 10 mM DTT incubation is used as positive control of caspase activation to verify that purified monomers are functional and inducible (*see* **Note 9**). For the negative control, add caspase-1 inhibitor Ac-YVAD-fmk (2 µM).
3. Add 4 µL/well sterile DDW and incubate the plate at 37 °C for 30 min. While the plate is incubating, prepare the caspase-1 fluorescent substrate Ac-WEHD-AMC (20 µM/well) and set the fluorometer in kinetic mode for 1 h with appropriate filters and wavelengths.
4. Add 2 µL/well of the Ac-WEHD-AMC.
5. Monitor the release of AMC product continuously by spectrofluorometry in kinetic mode over 1 h at 25–27 °C. Calculate the catalysis rates using the initial slope and express caspase activity as the change in fluorescence over time derived from the linear phase of reactions. A significant difference of caspase activity between 1 and 10 mM DTT incubations should be obtained when the amount of enzyme is kept constant (*see* **Note 9**).
6. To test NLRP1-dependent activation, determine the amount of pro-caspase-1 monomers showing (a) the lowest activity

background with 1 mM DTT incubation, and (b) the highest activity increase in the presence of 10 mM DTT. This concentration may vary based on instrument specificities available for the assay (sensitivity, acquisition rate, etc).

Similar to pro-caspase-1 monomers, perform a calibration experiment to determine the appropriate amount of recombinant NLRP1 to detect specific MDP-LD/ATP/Mg^{2+}-dependent caspase activation:

7. Use the optimal amount of pro-caspase-1 determined in the "calibration experiment," and perform a serial dilution of GST- or His-NLRP1 prepared in Buffer C. Test increasing amounts of pro-caspase-1 starting from the appropriate concentration determined in **step 6** of Subheading 3.3.1. Adjust the protein concentration by adding BSA so that the amount of protein in the assay is constant in the serial dilutions of NLRP1. Prepare 2 μL of the mix.
8. Mix recombinant NLRP1 (2 μL) with pro-caspase-1 monomers (2 μL) and set in the plate (4 μL/well).
9. Add 2 μL/well of 2× caspase-1 buffer.
10. Prepare two sets of NLR ligands: 1. no MDP-LD but Mg^{2+} (0.5 mM) and ATP (0.25 mM); 2. MDP-LD (0.1 μg/mL), Mg^{2+} (0.5 mM) and ATP (0.25 mM). These reagents should be in a stock concentration. Add 2 μL/well of the mix.
11. Incubate the plate at 37 °C for 30 min.
12. Prepare the caspase-1 fluorescent substrate Ac-WEHD-AMC (20 μM/well from 1 mM stock solution). Set the fluorometer to kinetic mode with appropriate filters, excitation and emission wavelengths.
13. Add 2 μL/well of the Ac-WEHD-AMC.
14. Monitor the release of AMC product continuously by spectrofluorometry in kinetic mode over 1 h at 25–27 °C (*see* **Note 15**).
15. Determine the amount of recombinant NLRP1 you will be working with (*see* **Note 16** and Fig. 1). As negative controls for NLRP1-dependent activation, use MDP-DD inactive enantiomer or γ -tri-DAP (NOD1 ligand) in the assay instead of active MDP-LD.

3.3.2 In Vitro ATP Binding to NLRP1

1. Prepare increasing concentrations of monomeric or aggregated recombinant NLRP1 (up to 1 μM) in Buffer C (*see* **Note 17**). Typically 8 μL/well of each dilution is added in a black flat bottom 96-well plate. Prepare triplicates.
2. Prepare 2 μL/well of a mix of NLRP1 ligands containing MDP-LD (1 μM), FITC-conjugated ATP analog (10 nM) and $MgCl_2$ (0.5 mM), with or without various concentrations of unlabeled ATP. Prepare triplicates.

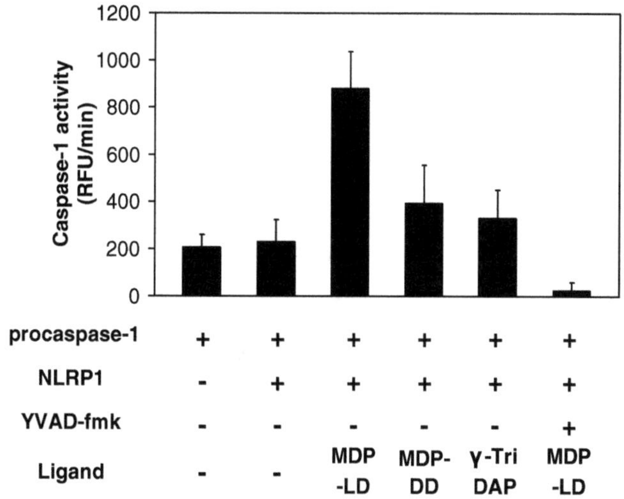

Fig. 1 GST-6His-NLRP1 (8.5 nM) and 6His-pro-caspase-1 (8.5 nM) were incubated either with MDP-LD or MDP-DD (0.1 μg/mL), or with γ Tri-DAP (0.1 μg/mL), in the presence of Mg^{2+} (0.5 mM) and ATP (0.25 mM), with or without caspase-1 inhibitor YVAD-fmk (2 μM). Caspase-1 activity was measured using Ac-WEHD-AMC substrate (20 μM) and expressed as relative fluorescence units (RFU)/min (mean ± SD; $n = 3$)

3. Add the mix of ligands to NLRP1 dilutions in the plate and incubate for 15 min on ice.
4. After incubation, read the plate at room temperature with the spectrofluorometer set in fluorescence polarization mode. ATP binding is quantified by measuring milliPolars (mPs) (*see* **Note 18**).
5. The FITC-ATP bound portion that is not competed by an excess of unlabeled ATP is considered nonspecific. A typical Protein/Ligand binding hyperbolic curve should be obtained by plotting mPs = f(NLRP1) (*see* Figs. 2 and 3) [14]. ATP affinity can be determined from this plot by calculating the disassociation constant of equilibrium K_D [14, 16].

4 Notes

1. The human NLRP1 cDNA transfected in Sf9 cells is a splice variant lacking the auto-cleavage site. Thus, cleaved forms are not observed. Therefore, avoiding auto-proteolysis may be important when isolating NLRP1 monomers.
2. Sf9 cells used to prepare recombinant baculoviruses, i.e., during co-transfections, or protein expression, should be used fresh and at the lowest passage possible. We found that either TNM-FH (Pharmingen) or Sf900III (Invitrogen) insect

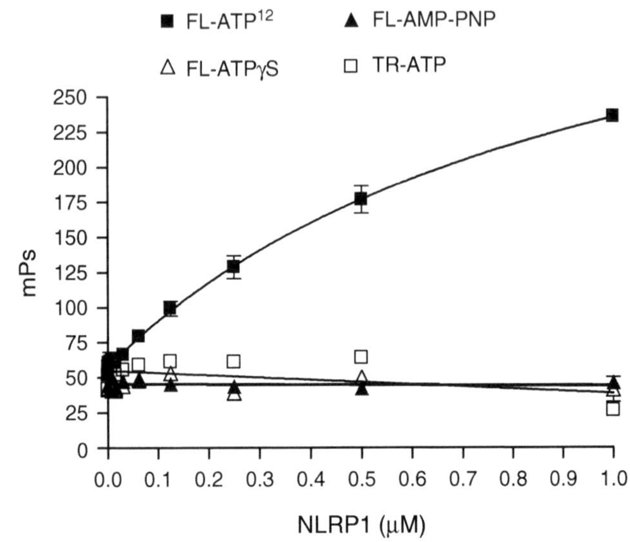

Fig. 2 GST-6His-NLRP1 was incubated with various FL-conjugated ATP analogs (10 nM) and Mg^{2+} (0.5 mM) for 5 min on ice in the presence of MDP-LD (20 μg/mL). ATP binding was analyzed by FPA (mean ± SD; $n = 3$ milliPolars [mPs]). NLRP1 binds only to FL-ATP where fluorophore conjugation is to the adenine group (FL-ATP[12])

medium gave similar results. High quality and concentrated recombinant DNA should be used during co-transfection experiments to yield a good rate or recombinant viruses and protein expression. Recombinant DNA to transfect should be properly purified, with 260/280 nm: ~1.9 and highly concentrated to allow a low mix volume during co-transfection.

3. Make sure that the entire plate is covered with Transfection Buffer A, to prevent the cells from drying out.
4. During transfection efficiency trials, we noticed the Baculogold:Transfer vector ratio should be 1:10.
5. Non-transfected cells should look healthy, well attached and at 90–100 % confluency. The transfected cells will be smaller and look partially lysed as they are on late-stage viral cycles and should release recombinant baculoviruses.
6. Cells should be swollen and much bigger than non-infected cells. If the infection was not efficient, then the cells will look well rounded, brighter, and smaller under the microscope.
7. For protein expression, infect cells with fresh recombinant viruses (not frozen) in SUP#3. Do not use viruses obtained from more virus amplification cycles, this will increase the rate of truncated forms of proteins. With our reagents generated, we observed that infecting Sf9 cells at MOI5 during 48 h was the best compromise between protein quality (not cleaved and soluble), and yield.

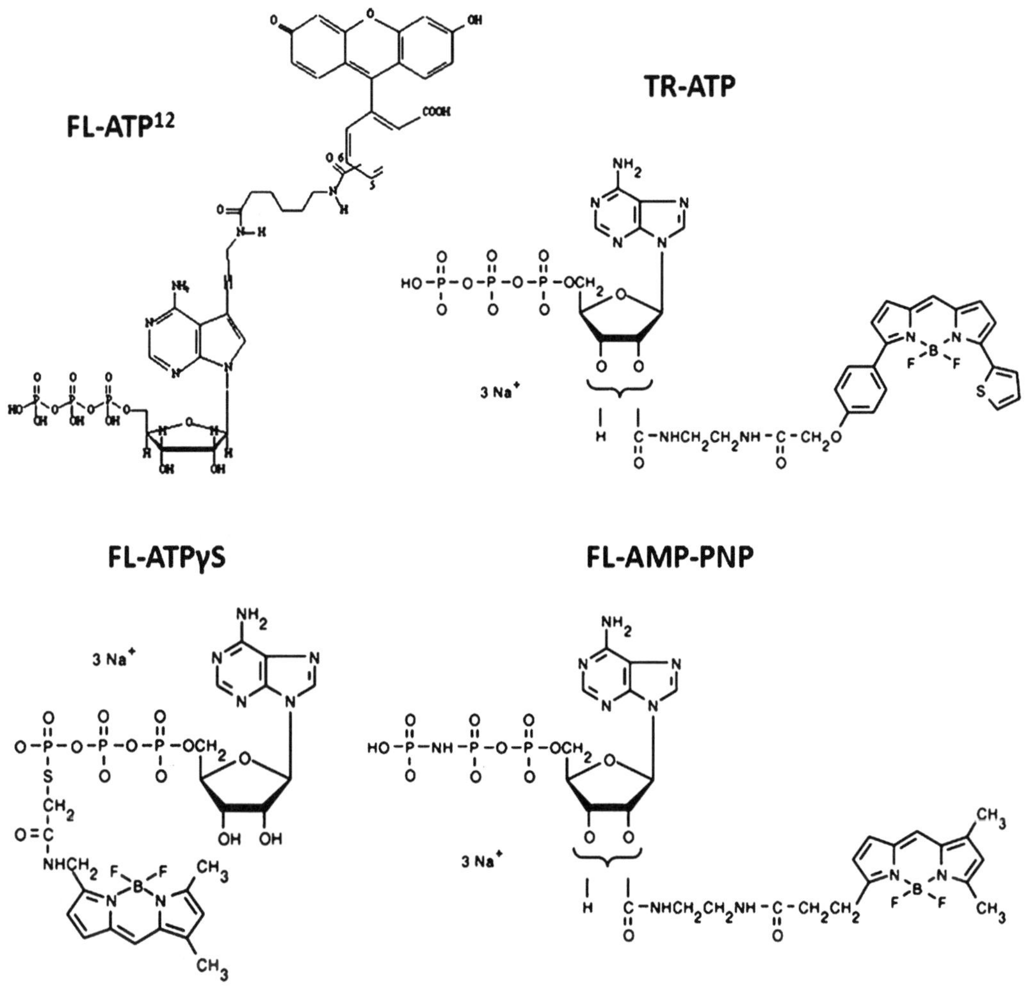

Fig. 3 Structures of the various fluorophore conjugated-ATP analogs used for Fluorescence Polarization Assays. The structures of fluorophore-conjugated ATP analogs are presented. Fluorophores (FL or Texas Red) are covalently bound to different locations on the ATP molecule, including 2′-(or-3′) ribose (Texas Red-ATP and FL-AMP-PNP), or γ-phosphate (FL-ATP-γ-S), or adenine group (FL-ATP12). FL-ATP12 was purchased from PerkinElmer, and the others from Invitrogen

8. It is better to start the purification procedure from a fresh cell pellet, rather than frozen. NLRP1 is a large protein (~180 kDa) highly sensitive to protease degradation. During the purification procedure, much attention should be taken to work with very clean materials, at least ethanol washed if not sterile. Otherwise, NLRP1 may be activated by residual bacterial components and may oligomerize. Therefore, filter all buffers and avoid any contamination by using autoclaved glassware and sterile tissue culture tubes. In addition, the Superdex 200 column for size exclusion chromatography should be stored in $H_2O^{+}NaO_3$. All buffers should be prepared fresh and filtered. Inject proteins into the AKTA purifier with sterile syringes.

9. Incubation of pro-caspase-1 monomers with 10 mM DTT releases caspase activity. High DTT allows maintaining reduced Cys^{285} residue in the catalytic site that is essential for enzyme auto-processing activity. The difference of caspase activity between 1 and 10 mM DTT is further observed following increased concentration of pro-caspase-1 in the assay, as higher amount of available caspase-1 monomers displaces the equilibrium toward dimer formation and enhances auto-processing. Cleaved caspase fragments can be detected as a result of incubation with 10 mM DTT.
10. Control that protein peaks, monitored with absorbance (measured at 280 nm) are well separated from each other as expected from the manufacturer manual.
11. The peak should be quite large, not sharp but stretched. To note, the main part of the protein aggregates and the amount of monomers is low. Do not use Thrombin enzyme to remove the GST tag as it cleaves NLRP1 at several sites.
12. 50 μg/mL is sufficient for the NLRP1 inflammasome assay in a 96-well plate. Do not concentrate the pro-caspase-1 monomers as this will displace the equilibrium toward dimer formation and result in subsequent auto-processing. The p45 monomeric form is the inactive zymogen, irrespective to dimers $(p45)_2$ and fully active cleaved hetero-dimer $(p10/p20)_2$ [14, 17].
13. Very sensitive protein band detection can be obtained by using SYPRO Ruby fluorescent gel staining, and a UV laser scanner. With this methodology, recombinant proteins obtained should be fairly pure (>85 % purity) as determined by densitometry analysis. By Western blot, GST-6His-NLRP1 is detectable as a single band at size ~200 kDa using monoclonal anti-GST antibody 1/1,000 dilution (Sigma), and 6His-pro-caspase-1 at size 50 kDa using monoclonal anti-caspase-1 antibody 1/1,000 dilution (Imgenex, recognizes full length and p10 fragment).
14. Recovery of GST- and His6-NLRP1 monomers are low and may fluctuate from one preparation to another, depending on the stability and efficiency of virus recombination, quality of Sf9 cells, and the purification procedure. Therefore, calibration experiments are mandatory for each batch produced, to characterize protein activity. Due to the low yield, several rounds of protein expression/purification using fresh viruses are needed to perform large volume experiments.
15. Rates of catalysis are calculated using the release of fluorescence over time during the linear phase of reactions. This phase is typically obtained during the first or the last 30 min following addition of Ac-WEHD-AMC (depending on the amount of enzyme used in the assay). Caspase activity is expressed as RFU/min.

16. Typically, the best condition combines several parameters: (a) addition of recombinant NLRP1 to pro-caspase-1 monomers should not induce a major increase of caspase-1 activity compared to pro-caspase-1 alone (in the absence of ligands); (b) MDP-LD, ATP and $MgCl_2$ are all essential to induce caspase activation in the presence of NLPR1 monomers and pro-caspase-1 monomers; and (c) caspase activity monitored in the presence of NLRP1/pro-caspase-1 and MDP-LD/ATP/Mg^{2+} should be at least five times greater vs. no ligands (as shown in [14, 16] [*see* Fig. 1]).
17. We observed that aggregated GST- or His6-NLRP1 isolated by size exclusion chromatography from the void volume (>670 kDa_{app}) is also competent for ATP binding using the Fluorescent Polarization Assay.
18. Fluorescence polarization measurements provide information on molecular orientation, mobility, and the processes that modulate them, including protein–ligand interactions. Experimentally, the degree of polarization is determined from measurements of fluorescence intensities parallel and perpendicular with respect to the plane of linearly polarized excitation light, and is expressed in terms of fluorescence polarization (milliPolars = mPs). The spectrofluorometer measures both fluorescence intensities and calculates mPs that are used as readouts to quantify ATP binding. When fluorescent ATP binds to NLRP1, it polarizes, which translates to high mP values (as observed in Fig. 2 with Fluorescein-12-ATP).

References

1. Martinon F, Burns K, Tschopp J (2002) The inflammasome: a molecular platform triggering activation of inflammatory caspases and processing of proIL-beta. Mol Cell 10(2):417–426
2. Hornung V, Ablasser A, Charrel-Dennis M, Bauernfeind F, Horvath G, Caffrey DR, Latz E, Fitzgerald KA (2009) AIM2 recognizes cytosolic dsDNA and forms a caspase-1-activating inflammasome with ASC. Nature 458(7237):514–518. doi:nature07725 [pii], 10.1038/nature07725
3. Schroder K, Tschopp J (2010) The inflammasomes. Cell 140(6):821–832. doi:S0092-8674(10)00075-9 [pii], 10.1016/j.cell.2010.01.040
4. Davis BK, Wen H, Ting JP (2011) The inflammasome NLRs in immunity, inflammation, and associated diseases. Annu Rev Immunol 29:707–735. doi:10.1146/annurev-immunol-031210-101405
5. Saleh M (2011) The machinery of Nod-like receptors: refining the paths to immunity and cell death. Immunol Rev 243(1):235–246. doi:10.1111/j.1600-065X.2011.01045.x
6. De Nardo D, Latz E (2011) NLRP3 inflammasomes link inflammation and metabolic disease. Trends Immunol 32(8):373–379. doi:doi:S1471-4906(11)00082-2 [pii], 10.1016/j.it.2011.05.004
7. Doyle SL, Campbell M, Ozaki E, Salomon RG, Mori A, Kenna PF, Farrar GJ, Kiang AS, Humphries MM, Lavelle EC, O'Neill LA, Hollyfield JG, Humphries P (2012) NLRP3 has a protective role in age-related macular degeneration through the induction of IL-18 by drusen components. Nat Med 18:791–798. doi:10.1038/nm.2717, nm.2717 [pii]
8. Salvesen GS (2002) Caspases and apoptosis. Essays Biochem 38:9–19
9. Nagata S (1997) Apoptosis by death factor. Cell 88(3):355–365. doi:S0092-8674(00)81874-7 [pii]
10. Reed JC, Doctor K, Rojas A, Zapata JM, Stehlik C, Fiorentino L, Damiano J, Roth W,

Matsuzawa S, Newman R, Takayama S, Marusawa H, Xu F, Salvesen G, Godzik A (2003) Comparative analysis of apoptosis and inflammation genes of mice and humans. Genome Res 13(6B):1376–1388

11. Jin Y, Mailloux CM, Gowan K, Riccardi SL, LaBerge G, Bennett DC, Fain PR, Spritz RA (2007) NALP1 in vitiligo-associated multiple autoimmune disease. N Engl J Med 356(12):1216–1225
12. D'Osualdo A, Weichenberger CX, Wagner RN, Godzik A, Wooley J, Reed JC (2011) CARD8 and NLRP1 undergo autoproteolytic processing through a ZU5-like domain. PLoS One 6(11):e27396. doi:10.1371/journal.pone.0027396, PONE-D-11-15100 [pii]
13. Tschopp J, Martinon F, Burns K (2003) NALPs: a novel protein family involved in inflammation. Nat Rev Mol Cell Biol 4(2):95–104
14. Faustin B, Lartigue L, Bruey JM, Luciano F, Sergienko E, Bailly-Maitre B, Volkmann N, Hanein D, Rouiller I, Reed JC (2007) Reconstituted NALP1 inflammasome reveals two-step mechanism of caspase-1 activation. Mol Cell 25(5):713–724
15. Bruey JM, Bruey-Sedano N, Luciano F, Zhai D, Balpai R, Xu C, Kress CL, Bailly-Maitre B, Li X, Osterman A, Matsuzawa S, Terskikh AV, Faustin B, Reed JC (2007) Bcl-2 and Bcl-XL regulate proinflammatory caspase-1 activation by interaction with NALP1. Cell 129(1):45–56
16. Faustin B, Chen Y, Zhai D, Le Negrate G, Lartigue L, Satterthwait A, Reed JC (2009) Mechanism of Bcl-2 and Bcl-X(L) inhibition of NLRP1 inflammasome: loop domain-dependent suppression of ATP binding and oligomerization. Proc Natl Acad Sci U S A 106(10):3935–3940. doi:0809414106 [pii], 10.1073/pnas.0809414106
17. Thornberry NA, Bull HG, Calaycay JR, Chapman KT, Howard AD, Kostura MJ, Miller DK, Molineaux SM, Weidner JR, Aunins J et al (1992) A novel heterodimeric cysteine protease is required for interleukin-1 beta processing in monocytes. Nature 356(6372):768–774

Chapter 12

Assessing ATP Binding and Hydrolysis by NLR Proteins

Jinyao Mo and Joseph A. Duncan

Abstract

Nucleotide-binding and leucine rich repeat domain-containing proteins (NLRs) are central to the formation of many inflammasome complexes. Several inflammasome forming NLR proteins are known to be ATPases, but the nucleotide binding specificity of many remains to be characterized. The oligomerization of NLR proteins and assembly of inflammasomes require the ATP (or other nucleotide) binding activity of the NLR proteins. Quantitative and qualitative studies of the nucleotide binding properties of these proteins are useful tools in studying the regulation of inflammasome activity, and are outlined in this Chapter.

Key words Nucleotide binding, Leucine rich repeat domain-containing protein, ATPase

1 Introduction

NLR proteins were the first class of proteins recognized to function in assembling caspase-1 activating complexes known as inflammasomes [1, 2]. Since that original observation, other proteins have also been found to assemble inflammasome complexes [3]. The assembly of the inflammasome depends on the regulated oligomerization of the complex's primary scaffolding protein. In reported studies of NLR protein activity, the oligomerization of NLR proteins is dependent on their nucleotide binding activity [4, 5].

NLRs have a conserved tripartite primary structure consisting of an amino-terminal effector domain, central nucleotide-binding domain, and carboxyl-terminal series of leucine rich repeats [6]. These structural elements are similar to a subfamily of proteins encoded by plant disease resistance genes (R genes), which generally have an alternative amino-terminal domain motif, and to the mammalian Apoptotic Protease Activating Factor-1 (APAF-1), which carries a series of carboxyl-terminal WD-40 motifs rather than leucine-rich repeats. The central, nucleotide-binding domain is homologous to the AAA-ATPase superfamily. Similar to observations with APAF-1 and plant R genes, nucleotide binding and/or hydrolysis appear to play an important regulatory role in NLR protein function [7–10].

Christine M. De Nardo and Eicke Latz (eds.), *The Inflammasome: Methods and Protocols*, Methods in Molecular Biology, vol. 1040, DOI 10.1007/978-1-62703-523-1_12, © Springer Science+Business Media New York 2013

Many NLR-encoding genes have strong genetic association with human diseases [11]. The importance of nucleotide binding in NLR protein function was further supported by the discovery that a number of highly penetrant, inherited, autoinflammatory disorders are associated with missense mutations in the nucleotide binding domains of these genes [12–17]. In particular, mutations in the gene encoding NLRP3 that lead to alterations in the nucleotide binding domain are associated with three dominantly inherited periodic fevers: Familial Cold Autoinflammatory Syndrome (FCAS), Muckle–Wells Syndrome (MWS), and Neonatal Onset Multisystemic Autoinflammatory Disease (NOMID) [16, 18–21]. Patients with any of the three syndromes have elevated IL-1β levels in their serum, which is thought to result from constitutive assembly of the NLRP3 inflammasome [22, 23].

The role of nucleotide binding in NLR protein function has been explored in a handful of NLR proteins. These studies have demonstrated that most NLR proteins appear to be ATP or dATP binding proteins with two notable exceptions: (1) the MHC class II transcriptional activator (CIITA) was found to bind to GTP and its GTP binding activity controls trafficking of the protein into the nucleus and (2) all triphosphate nucleotides were found to support in vitro inflammasome assembly by NLRP1 [5, 24]. Direct nucleotide binding studies by NLR proteins requires a source of purified protein. To date, there are no published studies in which a native NLR protein has been purified from an endogenous source using affinity or conventional chromatography. Isolated NLR protein nucleotide-binding domains have been expressed as fusion proteins with Glutathione-*S*-Transferase (GST) and Maltose Binding Protein (MBP) in *E. coli* expression systems and can be purified to near homogeneity [25, 26]. Further, the purification of affinity-tagged, recombinant, full-length NLR proteins has now been reported using a variety of mammalian and insect cell expression systems [4, 5, 27, 28]. Both isolated nucleotide-binding domains and purified full length proteins can be used for studies of nucleotide binding and hydrolysis. We show the partial purification of NLRP3 using a tandem affinity tag purification strategy that we recently described for NOD2 (*see* Fig. 1) [28].

Many assays for studying protein–nucleotide interactions are now available, where nucleotide ligands can be modified with either radioactive isotopes (most commonly ^{3}H, ^{32}P, and ^{35}S) without altering their chemical structure, or with small molecule labels (biotin or fluorescent moieties) through chemical linkers. Fluorescence anisotropy, which measures changes in molecular tumbling that result from a fast tumbling small molecule (like a fluorescently labeled nucleotide analog) binding to a large slower tumbling molecule (like a protein), has previously been used to characterize nucleotide binding by NLRP1 and NOD1 [5, 27]. We have utilized radiolabeled nucleotide ligands to study

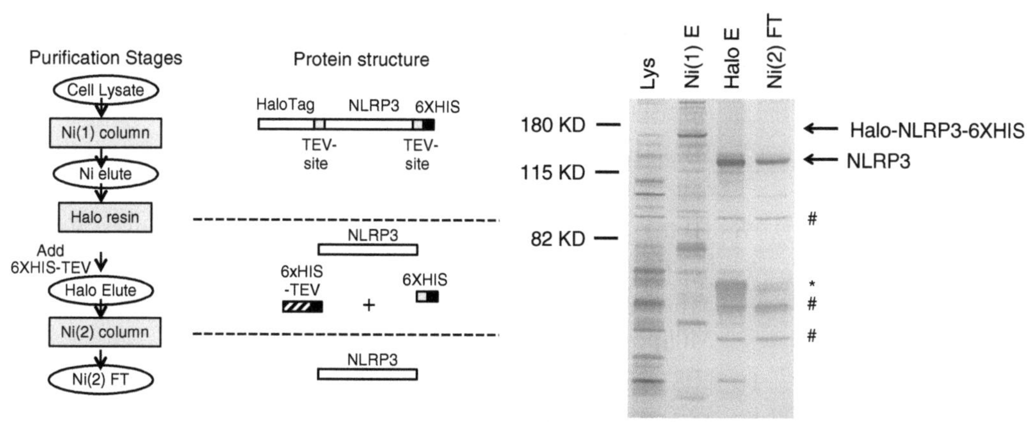

Fig. 1 Partial purification of recombinant NLRP3 for ATP binding studies. Sf9 cells were infected with recombinant baculovirus expressing Halo-NLRP3-6xHIS, as previously described for expression of recombinant NOD2 [28]. The cells were lysed (*Lys*) and Halo-NLRP3-6xHIS was enriched by elution from an immobilized metal chromatography resin (*Ni(1) E*). The enriched Halo-NLRP3-6xHIS was captured using Halo-ligand chromatography (Promega) and NLRP3 free of the Halo and 6xHIS tags was eluted by cleavage with recombinant 6xHIS-TEV protease (*Halo E*). TEV protease was removed from the preparation by passing the recombinant NLRP3-containing through a second immobilized metal affinity chromatography resin (*Ni(2) FT*). 25 ng of protein from each stage of purification (*lanes* as indicated) was subjected to SDS-PAGE and total protein was visualized with silver staining. The location of Halo-NLRP3-6xHIS and NLRP3 are marked with *arrows*. The location of 6xHIS-TEV protease is indicated by an *asterisk* and contaminating proteins in the preparation are noted by the *hash symbol*

nucleotide-binding properties of NLR proteins, including NLRP3, NLRP12, and NOD2 [4, 26, 28]. Herein, we detail the methodology for three assays: (1) binding of a non-hydrolyzable ATP analog, Adenosine-5′-*O*-(3-thiotriphosphate) or ATPγS; (2) binding of a photo-activated ATP analog; and (3) steady-state ATPase activity assessed by measuring release of the terminal orthophosphate from ATP.

2 Materials

There is no consensus, generalizable methodology for NLR protein isolation. Expression of NLR proteins fused with two different affinity tags in insect cell culture using recombinant baculovirus followed by tandem affinity purification has been particularly successful in our hands [4, 28] (*see* **Note 1**).

2.1 Measurement of Adenosine-5′-O-(3-thiotriphosphate) (ATPγS) Binding by NLR Proteins

1. NLR protein storage buffer: 50 mM NaHEPES, pH 7.5, 500 mM NaCl, 1 mM DTT, and 0.1 % (w:v) CHAPS.
2. House vacuum system or vacuum pump (*see* **Note 2**).
3. Laboratory bench and fume hood designated for use of radioactivity materials (*see* **Note 3**).

4. Portable heat lamp.
5. Liquid scintillation counter.
6. Adjustable temperature water bath.
7. 10 place vacuum manifold with 25 mm filter holders (Hoefer, San Francisco, CA).
8. 10 mL capacity adjustable repeating syringe dispenser: Varipet Syringe Dispenser (Bel-Art Products, Wayne, NJ).
9. MF-Membrane HA filter (0.45 μm pore size, 24 mm diameter, Millipore).
10. [^{35}S]-ATPγS (1,250 Ci/mmol, PerkinElmer, Waltham, MA).
11. 5 mL plastic liquid scintillation vials.
12. ScintiSafe Econo 2 cocktail (Fisher Scientific) or other liquid scintillation fluid.
13. 2× ATP binding buffer: 100 mM Tris–HCl, pH 7.5, 300 mM NaCl, 4 mM DTT, 50 mM $MgCl_2$, and 10 % Glycerol (*see* **Note 4**).
14. Wash buffer: 20 mM Tris–HCl, pH 7.5, 120 mM NaCl, 25 mM $MgCl_2$, and 0.05 % Triton X-100.
15. 9×9 in. pyrex pan.
16. 15 mL conical screw cap tubes.
17. 1.5 mL microfuge tubes.
18. Aluminum foil.

2.2 Photo-affinity Labeling of NLR Proteins with 8-Azido-ATP Analogs

1. UV light (254 nm): UV cross-linking apparatus (UV StrataLinker 1800, Stratagene).
2. Savant SpeedVac (Thermo Scientific) or other centrifugal evaporator system.
3. FluorChemE multimode, CCD based imaging system (Protein Simple, Santa Clara, CA).
4. 8-N_3-ATP[γ]biotinpentylamine (Affinity Photoprobes, Lexington, KY).
5. Cross-linking buffer: 50 mM Tris–HCl, pH 7.5, 0.1 % CHAPS, 150 mM KCl, and 5 % glycerol.
6. 1× TBST: 20 mM Tris–HCl, pH 7.5, 150 mM NaCl, 0.5 % (v:v) Tween-20.
7. Blocking buffer: 5 % (w:v) dry milk in 1× TBST.
8. Streptavidin-Horse Radish Peroxidase conjugate.
9. SeeBlue® Plus2 Protein size markers (Invitrogen, Life technologies) or other protein size markers.
10. Precast NuPAGE® 4–12 % (1.0 mm) Bis-Tris gel and NuPAGE® MOPS running buffer (Invitrogen, Life Technologies).

11. NuPAGE® SDS-PAGE gel running apparatus and power supply (Invitrogen, Life Technologies).
12. Nitrocellulose iBlot® transfer stacks and iBlot® gel transfer device (Invitrogen, Life Technologies).
13. NuPAGE® LDS sample buffer.
14. NuPAGE® sample reducing agent.
15. SuperSignal West Pico Chemiluminescent Substrate (Pierce, Thermo Scientific).

2.3 Measurement of Steady State ATPase Activity of NLR Proteins

1. Laboratory bench and fume hood designated for use of radioactivity materials (*see* **Note 3**).
2. Activated charcoal (Sigma-Aldrich).
3. [γ-^{32}P]-ATP, 3,000 Ci/mmol (PerkinElmer).
4. Liquid scintillation counter.
5. 5 mL plastic liquid scintillation vials (RPI, Mt. Prospect, Ill).
6. Scintillation liquid: ScintiSafeEcono 2 LSC-cocktail (Fisher Scientific).
7. Water bath.
8. Microcentrifuge.
9. 1.5 mL microfuge tubes.
10. 100 mL graduated cylinder.
11. Parafilm.
12. Acidic phosphate solution: 100 mM HCl/100 mM Na_2HPO_4. Weigh 13.7 g $Na_2PO_4{\cdot}7H_2O$, add to a beaker containing 400 mL dH_2O. Stir to dissolve the chemical and add 4.2 mL concentrated HCl to the solution. Continue to stir. Make up to 500 mL with dH_2O (*see* **Note 5**).
13. 1× ATP binding buffer specified in **step 13** of Subheading 2.1.

3 Methods

3.1 Measurement of Adenosine-5′-O-(3-thiotriphosphate) (ATPγS) Binding by NLR Proteins

In this assay, purified NLR protein mixtures of proteins containing NLR proteins are incubated with a radiolabeled, non-hydrolyzable analog of ATP, [^{35}S]-ATPγS. After the binding reaction is complete, the protein bound nucleotide is isolated from the free nucleotide by filtering the solution through a protein-binding nitrocellulose filter. The filter-associated label is then detected using liquid scintillation counting. Variations of this assay can be performed to assess the relative nucleotide binding stoichiometry of different proteins (or preparations of proteins), the relative preference of a particular NLR protein for a nucleotide base by including titrations of competitor nucleotides, or the relative rates of

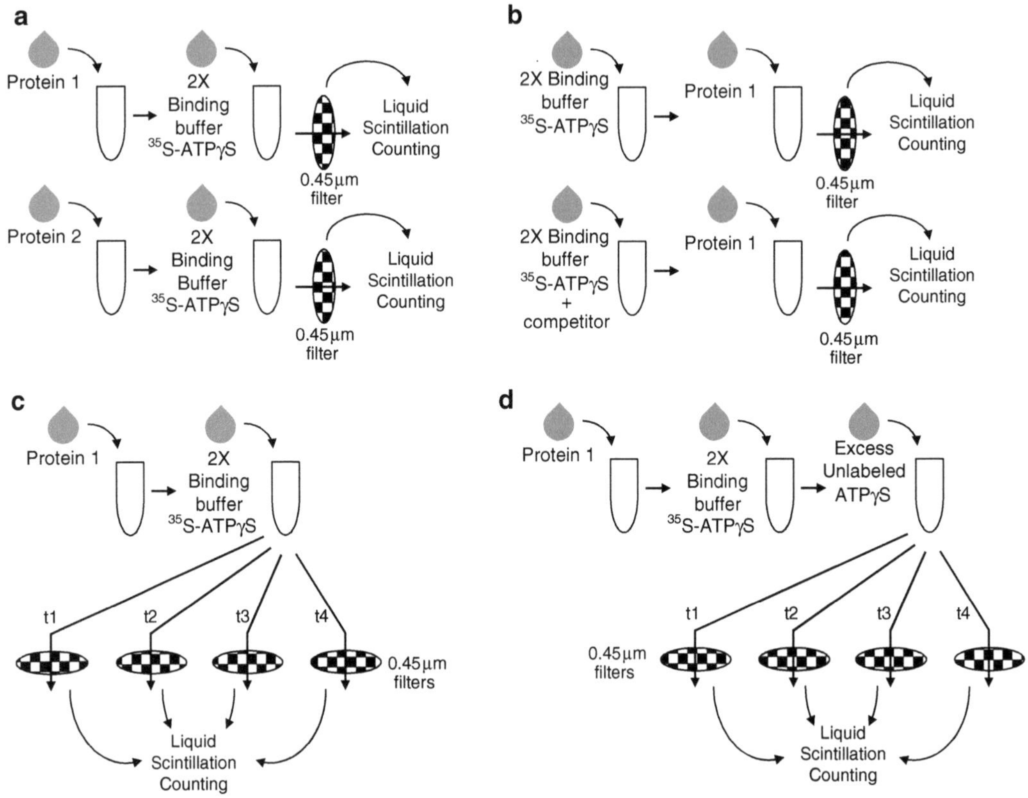

Fig. 2 Experimental outlines for studies of NLR proteins utilizing ATPγS filter binding method. (**a**) Comparison of ATP binding stoichiometry between protein preparations: binding reactions for each protein to be compared were set up in parallel and allowed to proceed to equilibrium prior to measurement. (**b**) Comparison of inhibitors (small molecule inhibitors, potential regulatory proteins, or competitive binding with different nucleotides) of NLR protein nucleotide binding: inhibitors were mixed with [^{35}S]-ATPγS prior to starting binding reaction. (**c**) Rate of ATPγS binding to NLR proteins: NLR protein(s) was added to [^{35}S]-ATPγS binding reaction and aliquots of the reaction were collected at specified time points and analyzed for bound [^{35}S]-ATPγS. (**d**) Rate of ATPγS disassociation from NLR proteins: the NLR protein was incubated with [^{35}S]-ATPγS until equilibrium was reached, an excess of unlabeled- ATPγS (usually 100–1,000-fold) was added, and NLRP3 bound [^{35}S]-ATPγS was measured at specified time points after addition of unlabeled nucleotide

association or dissociation of an NLR protein from ATPγS (*see* Fig. 2). Additionally, the optimal buffer pH, NaCl concentration, and Mg^{2+} ion concentration may vary from protein to protein. We recommend optimizing these when exploring ATPγS-binding activity of uncharacterized NLR proteins. The description of the assay that follows reflects the first of these assays, a comparison of relative nucleotide binding by different NLR containing protein preparations. The assay is carried out at a single concentration of ATPγS for a fixed time of binding (a time that has previously been demonstrated achieves maximum binding for NLRP3). In the described assay, we compare the ATPγS binding of NLRP3 at four stages of protein preparation (*see* Figs. 1 and 3).

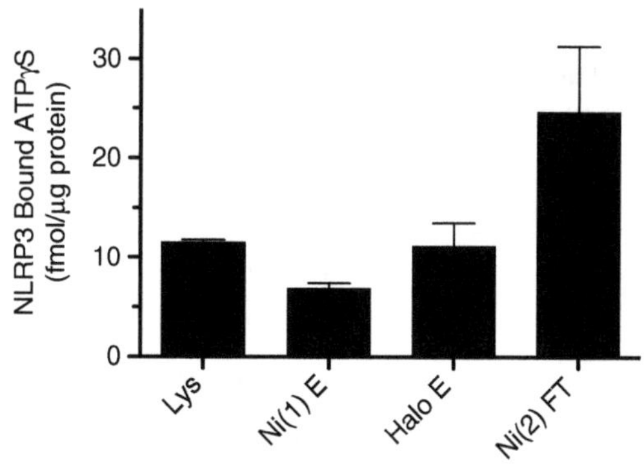

Fig. 3 [^{35}S]-ATPγS binding by recombinant NLRP3 at different stages of purification. 100 ng of protein from each stage of purification indicated in Fig. 1 was incubated with [^{35}S]-ATPγS for 60 min as indicated in the text. The amount of protein bound [^{35}S]-ATPγS was assessed for each preparation. The *bars* indicate mean of triplicate binding reactions and *error bars* show + S.D.

1. Determine the number of filters required for the assay (*see* **Note 6**).
2. Soak MF membrane filters in ~200 mL Wash buffer, in a 9 × 9 in. pyrex pan. To wet the membrane evenly, place filters in the pan and rock or shake slowly for at least 30 min at room temperature (*see* **Note 7**).

   While membranes are soaking, set up binding reactions:
3. Thaw and prepare 1 μL diluted [^{35}S]-ATPγS for each filtered reaction. Dilute [^{35}S]-ATPγS 1:16 in 2× ATP binding buffer. Use 1 μL of the diluted isotope in each 100 μL binding reaction, yielding a final concentration of 5 nM [^{35}S]-ATPγS in the reaction mix.
4. Bring water bath temperature to 30 °C.
5. Prepare 2× reaction mix (2× ATP binding buffer with [^{35}S]-ATPγS). Place 49 μL of 2× of binding buffer x (# of prepared filters) in a 15 mL conical screw cap tube (for a low number of binding reactions a 1.5 mL microfuge tube can be substituted) and add 1 μL x (# of prepared filters) of diluted [^{35}S]-ATPγS (*see* **Note 8**).
6. Label three 1.5 mL microfuge tubes for each sample to be tested and three additional tubes for a no protein control.
7. Add 50 μL of 2× binding buffer with [^{35}S]-ATPγS to each reaction tube.
8. Dilute protein samples being tested for assay. We typically use ~100 ng purified NLR protein for each binding reaction. We dilute purified protein preparations to 2 ng/μL in NLR Protein Storage Buffer (*see* **Note 6**).

Table 1
Experimental layout for testing [^{35}S]-ATPγS-binding capacity of recombinant NLRP3 at various stages of purification shown in Fig. 1

Tube number	Sample	Start time	Stop time
1	No protein (protein storage buffer)	0:0:0	1:00
2	No protein (protein storage buffer)	0:0:20	1:00:20
3	No protein (protein storage buffer)	0:0:40	1:00:40
4	NLRP3-expressing Sf-9 lysate	0:2:0	1:02
5	NLRP3-expressing Sf-9 lysate	0:2:20	1:02:20
6	NLRP3-expressing Sf-9 lysate	0:2:40	1:02:40
7	Ni^{2+} elution	0:4:0	1:04
8	Ni^{2+} elution	0:4:20	1:04:20
9	Ni^{2+} elution	0:4:40	1:04:40
10	Halo elution	0:6:0	1:06
11	Halo elution	0:6:20	1:06:20
12	Halo elution	0:6:40	1:06:40
13	2^{cd} Ni^{2+} FT	0:8:0	1:08
14	2^{cd} Ni^{2+} FT	0:8:20	1:08:20
15	2^{cd} Ni^{2+} FT	0:8:40	1:08:40

9. At timed intervals, add 50 μL protein solution (or Protein Storage Buffer) into the appropriate reaction tube, vortex and place reaction in a 30 °C water bath. Start all reactions for one sample type in 40 s intervals then wait 2 min and initiate the next reaction. Allow the binding reaction to proceed for 60 min. *See* Table 1 for starting and completion time example for assay run to generate Fig. 3 (*see* **Notes 9** and **10**).
10. Approximately 5 min before the end of the hour, turn on the vacuum for the filtration system and close the individual valves for each filter holder spot and prepare to process the reactions.
11. Lay three wet filter membranes on filter support grids of the vacuum manifold and place manifold chimneys in place over the filters (*see* **Note 11**).
12. Add 4 mL ice-cold 1× ATP binding buffer to filter each holder (*see* **Note 12**).
13. Add 100 μL (all) of the reaction mixture to the buffer in a filter holder and open the filter valve to begin the filtration of that sample.

14. Repeat the transfer of reaction mixtures to filter holders for all three reactions of a single sample at 20 s intervals.
15. Wash 3× by immediately adding 8 mL wash buffer to each holder after all buffer passes through the membrane using the Varipet Syringe Dispenser (for a total of 8 mL × 3) (*see* **Note 13**).
16. Turn off vacuum and remove filter membranes from the filtration system to a 20 × 20 cm sheet of aluminum foil.
17. Repeat **steps 11–16** of Subheading 3.1, for each reaction sample.
18. To determine the total cpm placed in each reaction and the specific activity (cpm/pmol) of the labeled [^{35}S]-ATPγS, run a set of three filters through **steps 11–16** skipping the addition of reaction mixture (**step 13**).
19. Dry membranes thoroughly using a heat lamp (*see* **Note 14**).
20. For the three filters prepared in **step 18**, pipette 2.5 μL of label containing 2× binding buffer (remaining from **step 5**) onto each filter and allow filters to completely dry. (The cpm detected on these filters represents 1/20 the total placed in the reaction).
21. Put the filters into 5 mL scintillation vials and add 5 mL scintillation liquid.
22. Count using a liquid scintillation counter using a program suitable for ^{35}S detection.

3.2 Photo-affinity Labeling of NLR Proteins with 8-Azido-ATP Analogs

Most NLR proteins are not efficiently purified to homogeneity and ATPγS binding stoichiometries tend to be a fraction of the expected number of molecules of NLR protein in the assay described above, thus, a secondary assay demonstrating that the NLR protein in the assay is responsible for the detected ATP binding is recommended. Photo-affinity labeling with an ATP-analog like 8-N_3-ATP[γ] biotinpentylamine allows visualization of the ATP-analog-associated protein after separation by SDS-PAGE. This method allows confirmation of ATP association with the expected protein in the preparation.

1. Place 2 μL of 50 μM 8-Nt_3-ATP[γ]biotinpentylamine into a 1.5 mL microfuge tube.
2. Place the tube in SpeedVac or other brand of vacuum assisted centrifugal concentrator, dry for 10 min (*see* **Note 15**).
3. Add 25 μL cross-linking buffer to the tube to dissolve 8-N_3-ATP[γ]biotinpentylamine.
4. Add 25 μL of protein-containing solution (between 300 and 600 ng of protein in the reaction) and mix thoroughly.
5. Irradiate samples for 3 min with UV light (254 nm) by placing open tubes in a UV cross-linking apparatus.

6. Add 15 μL 4× NuPAGE® LDS sample buffer (loading buffer) and 6.6 μL 10× NuPAGE® sample reducing agent to the reaction and incubate in a 95 °C heat block for 5 min (*see* **Note 16**).
7. Resolve the protein on NuPAGE® 4–12 % Bis-Tris gel run in MOPS running buffer according to the manufacturer's protocol.
8. Transfer the proteins to nitrocellulose according to the manufacturer's protocol.
9. Incubate the membrane in blocking buffer at room temperature for 1 h with shaking.
10. Dilute Streptavidin-Horse Radish Peroxidase conjugate 1:1,500 in blocking buffer (e.g., 10 μL in 15 mL).
11. Discard the blocking buffer on the membrane and replace it with the diluted Streptavidin-Horse Radish Peroxidase conjugate.
12. Incubate for 2 h at room temperature with shaking.
13. Discard the conjugate antibody and replace with 15 mL 1× TBST.
14. Incubate for 10 min at room temperature with shaking.
15. Repeat **steps 11–14** of Subheading 3.2, three additional times for a total of 4× 10 min washes.
16. Prepare SuperSignal West Pico chemiluminescent substrate by mixing 500 μL of each component.
17. Discard the last TBST wash and add 1 mL of chemiluminescent substrate to the surface of the membrane.
18. Allow detection reaction to proceed at room temperature for 3 min.
19. Visualize ATP-biotin bound proteins using FluorEChem CCD imaging system set for luminescent image capture or by X-ray film exposure.

3.3 Measurement of Steady State ATPase Activity of NLR Proteins

Measurement of ATP hydrolysis by NLR proteins can be assessed by measurement of released ^{32}P orthophosphate from [γ-^{32}P]-ATP substrate. The reaction is stopped by addition of a phosphoric acid/activated charcoal solution. ATP (and other nucleotides) bind to activated charcoal under these conditions while the orthophosphate product is free, allowing for separation of the substrate and product. Steady state hydrolysis is dependent on the affinity of the NLR protein for substrate, the concentration of NLR and substrate in the reaction, the rate of catalysis, and the rate of product/substrate exchange. By utilizing substrate concentrations several orders of magnitude greater than the Km of the protein for ATP, the rate limiting steps of the reaction are the catalytic rate and exchange rates. This reaction can initially be performed across a range of substrate concentrations to define the Km for any given NLR protein and to define the concentration of ATP to be used in future studies. The exchange rate is rapid for the only proteins of this family in which this has been measured and may be a point of regulation in the biological action of NLR proteins. Therefore, NLR protein steady state ATPase activity may be a useful assay in identifying the biochemical activity of NLR protein regulatory factors.

3.3.1 De-Fining Charcoal Solution

In this assay, activated charcoal is used to separate nucleotide-associated ^{32}P ([γ-^{32}P]-ATP), which binds activated charcoal, and released [^{32}P]-orthophosphate. The following procedure is used to prepare de-fined charcoal stop solution (*see* **Note 17**):

1. Place ~25 mL (packed powder volume) of activated charcoal powder in a 100 mL graduated cylinder.
2. Add ~75 mL (three times charcoal volume) of acidic phosphate solution. Cover the cylinder with parafilm and mix by inversion.
3. Let the cylinder sit at room temperature for 10 min to allow the charcoal to settle.
4. Carefully decant the liquid, which contains small charcoal particles (it is ok to lose a little charcoal powder).
5. Repeat **steps 2–4** of Subheading 3.3.1, five times. If small particles remain visible in the decanted solution, repeat an additional five times.
6. Estimate the volume of remaining charcoal powder in the cylinder and add acidic phosphate solution to the charcoal to reach the ratio of 3:1 (v/v). This charcoal stop solution is now ready for use.

3.3.2 Measuring ATPase Activity

1. Dilute [γ-^{32}P]-ATP 1:200 in 2× ATP binding buffer (*see* **Note 18**).
2. Make a 2× reaction buffer master mix: for each planned reaction time point, add 48 μL of 2× ATP binding buffer, 1 μL of diluted [γ-^{32}P]-ATP, and 1 μL of unlabeled ATP (100 μM, *see* **Note 19**).
3. Dilute proteins to be tested to appropriate concentration (40–160 ng/μL) in their normal storage buffer.
4. Prepare two 1.5 mL microfuge tubes with 50 μL of diluted protein/time point to be studied (250 μL for the assay conditions outlined in **step 2**).
5. Prepare two 1.5 mL microfuge tubes with 100 μL of protein storage buffer to be used as no protein control at the first and last time point.
6. Prepare stop reaction tubes (one for each time point to be studied, 14 for the experiment outlined above in **step 2**): 1.5 mL microfuge tube with 500 μL de-fined, phosphoric acid charcoal suspension (*see* **Note 20**).
7. Place stop reaction tubes in a rack with lids open.

8. Initiate each reaction at 30 s intervals by pipetting a volume of 2× reaction buffer that is equal to the volume of protein solution (or no protein buffer control), mix briefly in a vortex mixer, then place the reactions in a 30 °C water bath.
9. At $t = 5$ min, remove 100 μL of each reaction (protein containing reactions and no protein controls) and add immediately to one stop reaction tube containing acidified charcoal at 30 s intervals.
10. Close the stop reaction tube lid and mix with a single flick.
11. Once all 5 min time point samples have been stopped in charcoal, quickly vortex each sample.
12. Repeat **steps 9–11** for only the protein containing samples at the following times: $t = 15$, 25, and 35 min.
13. Repeat **steps 9–11** for both protein containing samples and no protein control at $t = 65$ min.
14. Centrifuge all stopped reactions at 13,000 × *g* for 5 min.
15. Remove 120 μL of each supernatant to a scintillation vial for counting.
16. Add 5 mL scintillation cocktail and count on a program appropriate for the detection of ^{32}P (total cpm released = 5 × cpm in scintillation vial).

3.3.3 Determining the Specific Activity of ATP in the Reaction

1. Prepare two stop reactions tubes (as described in **step 6** of Subheading 3.3.2) and add an additional 100 μL H_2O to each tube.
2. Centrifuge each tube in microfuge for 5 min at 13,000 × *g*.
3. Transfer 120 μL of supernatant from each tube into separate scintillation vials.
4. Dilute 2 μL of the reaction buffer master mix in 198 μL H_2O.
5. Spike 10 μL of diluted reaction buffer master mix from **step 4** of Subheading 3.3.3 (containing a total of 200 fmol ATP) into each scintillation vial, add 5 mL scintillation fluid, and count the vials on a program appropriate for detection of ^{32}P. The specific activity of [γ-^{32}P]-ATP is calculated as mean CPM from two vials/200 fmol.
6. Analyze data by plotting the release of ^{32}P against time (*see* **Note 21**).

4 Notes

1. The quantity of proteins utilized in these assays is based on our recent experience with the NLR protein NOD2 [28]. Depending on the system utilized to express and isolate the protein, there may be great variability in the quantity of protein available and the specific ATP binding of the proteins in those preparations.

2. The filtered solutions will contain radioactive waste, thus, it is important to ensure the vacuum system is appropriately isolated from the house vacuum using filters and waste flasks.
3. Utilize β emitting isotopes (^{35}S and ^{32}P) only in appropriately approved laboratory areas in compliance with your institution's regulations regarding safe handling of radioactivity. Utilize adequate plexiglass based shielding and appropriate personal safety protection.
4. We recently optimized these buffer conditions including pH, NaCl concentration, and $MgCl_2$ concentration for NOD2. For investigations, which will assess NLR proteins for which no conditions have been published, these parameters may require pilot testing to identify optimal conditions.
5. Use normal safety precautions when diluting strong acids.
6. Determine the number of filters required for the assay, as follows:
   (a) Three for counting total CPM in the binding reactions. +
   (b) Three for determining nonspecific filter binding (no protein control) +
   (c) 3× # of samples or time points to be tested.

   The assay depicted in Fig. 3 required 18 filters.
7. Filters may be soaked for more than 30 min if required. We have not noted a change in assay performance for filters soaked as long as 2 h.
8. For example, if testing binding of four protein preparations, as in Fig. 3, one would mix 882 μL of 2× ATP binding buffer + 18 μL diluted [^{35}S]-ATPγS.
9. If pilot experiments indicate the buffering and/or salt concentrations for the protein storage or optimized ATPγS binding are greatly different from those reported for this assay, the content of the 2× binding buffer may be adjusted to achieve the desired reaction conditions.
10. Precise timing of reaction initiation and stopping the reaction are not required when incubation time is extended beyond when maximum binding is achieved under the conditions of the assay (i.e., when equilibrium binding has been achieved).
11. To avoid leaking, you need to align membranes very well with the support grid, and ensure that when the chimney is placed, the entire circumference of the filter is in contact with the chimney.
12. Background ^{35}S binding to the filters is erratic if reaction mixtures are pipetted onto drying filters, it is important the reaction is diluted into a significant volume of wash then filtered immediately.
13. Keep the wash buffer bottle on ice in an ice bucket to keep the buffer chilled.

14. Variability in the moisture content of filters will lead to variable quenching during scintillation counting.
15. 8-N_3-ATP[γ]biotinpentylamine comes dissolved in methanol which might denature active NLR proteins if used without drying.
16. We routinely utilize the Invitrogen NuPAGE and iBlot systems for protein separation and immobilization on nitrocellulose. It is expected that other SDS-PAGE systems and electrophoretic transfer protocols can be used with similar efficiency.
17. Charcoal dust or fines are present in commercially available activated charcoal powder and will not readily pellet with centrifugation. This leads to suspended ^{32}P-labeled nucleotide contaminating the orthophosphate containing supernatant and unreasonably high background counts in the orthophosphate release assay.
18. This dilution typically yields enough cpm/reaction to allow reasonable detection (but this will vary with the age of the isotope). Liquid scintillation counting of the reaction mixture is required to determine the specific activity of the substrate in this assay.
19. We typically use a 5-point time course and run each reaction in duplicate. We also include a no protein control reaction run at the first and last time point, also run in duplicate. This experimental setup requires 700 μL of 2× reaction buffer master mix.
20. Mix charcoal suspension well by inverting the bottle several times before pipetting suspension or placing the bottle on a stir plate with constant stirring while pipetting contents. Also, cut off 2–3 mm of a 1 mL pipette tip to facilitate pipetting charcoal suspension.
21. If the data reasonably fits to a straight line, the slope of the line represents the steady rate of hydrolysis. If substrate is exhausted in this assay, the release of [^{32}P]-orthophosphate will slow down at later times as the concentration of substrate drops below ten times the Km. If this is the case, the data will not fit the linear equation well as the slope shifts towards horizontal.

Acknowledgment

This work was supported by the National Institutes of Health through grant AI088255 and the Burroughs Wellcome Fund Career Award for Medical Scientists.

References

1. Martinon F, Burns K, Tschopp J (2002) The inflammasome: a molecular platform triggering activation of inflammatory caspases and processing of proIL-beta. Mol Cell 10(2):417–426
2. Agostini L, Martinon F, Burns K, McDermott MF, Hawkins PN, Tschopp J (2004) NALP3 forms an IL-1beta-processing inflammasome with increased activity in Muckle-Wells autoinflammatory disorder. Immunity 20(3): 319–325
3. Hornung V, Ablasser A, Charrel-Dennis M, Bauernfeind F, Horvath G, Caffrey DR, Latz E, Fitzgerald KA (2009) AIM2 recognizes cytosolic dsDNA and forms a caspase-1-activating inflammasome with ASC. Nature 458(7237):514–518. doi:10.1038/nature07725
4. Duncan JA, Bergstralh DT, Wang Y, Willingham SB, Ye Z, Zimmermann AG, Ting JP (2007) Cryopyrin/NALP3 binds ATP/dATP, is an ATPase, and requires ATP binding to mediate inflammatory signaling. Proc Natl Acad Sci U S A 104(19):8041–8046
5. Faustin B, Lartigue L, Bruey JM, Luciano F, Sergienko E, Bailly-Maitre B, Volkmann N, Hanein D, Rouiller I, Reed JC (2007) Reconstituted NALP1 inflammasome reveals two-step mechanism of caspase-1 activation. Mol Cell 25(5):713–724
6. Ting JP, Lovering RC, Alnemri ES, Bertin J, Boss JM, Davis BK, Flavell RA, Girardin SE, Godzik A, Harton JA, Hoffman HM, Hugot JP, Inohara N, Mackenzie A, Maltais LJ, Nunez G, Ogura Y, Otten LA, Philpott D, Reed JC, Reith W, Schreiber S, Steimle V, Ward PA (2008) The NLR gene family: a standard nomenclature. Immunity 28(3):285–287
7. Genini D, Budihardjo I, Plunkett W, Wang X, Carrera CJ, Cottam HB, Carson DA, Leoni LM (2000) Nucleotide requirements for the in vitro activation of the apoptosis protein-activating factor-1-mediated caspase pathway. J Biol Chem 275(1):29–34
8. Jiang X, Wang X (2000) Cytochrome c promotes caspase-9 activation by inducing nucleotide binding to Apaf-1. J Biol Chem 275(40):31199–31203. doi:10.1074/jbc.C000405200
9. Tameling WI, Elzinga SD, Darmin PS, Vossen JH, Takken FL, Haring MA, Cornelissen BJ (2002) The tomato R gene products I-2 and MI-1 are functional ATP binding proteins with ATPase activity. Plant Cell 14(11): 2929–2939
10. Tameling WI, Vossen JH, Albrecht M, Lengauer T, Berden JA, Haring MA, Cornelissen BJ, Takken FL (2006) Mutations in the NB-ARC domain of I-2 that impair ATP hydrolysis cause autoactivation. Plant Physiol 140(4):1233–1245
11. Ting JP, Kastner DL, Hoffman HM (2006) CATERPILLERs, pyrin and hereditary immunological disorders. Nat Rev Immunol 6(3):183–195
12. Hugot JP, Chamaillard M, Zouali H, Lesage S, Cezard JP, Belaiche J, Almer S, Tysk C, O'Morain CA, Gassull M, Binder V, Finkel Y, Cortot A, Modigliani R, Laurent-Puig P, Gower-Rousseau C, Macry J, Colombel JF, Sahbatou M, Thomas G (2001) Association of NOD2 leucine-rich repeat variants with susceptibility to Crohn's disease. Nature 411(6837):599–603
13. Miceli-Richard C, Lesage S, Rybojad M, Prieur AM, Manouvrier-Hanu S, Hafner R, Chamaillard M, Zouali H, Thomas G, Hugot JP (2001) CARD15 mutations in Blau syndrome. Nat Genet 29(1):19–20
14. Ogura Y, Bonen DK, Inohara N, Nicolae DL, Chen FF, Ramos R, Britton H, Moran T, Karaliuskas R, Duerr RH, Achkar JP, Brant SR, Bayless TM, Kirschner BS, Hanauer SB, Nunez G, Cho JH (2001) A frameshift mutation in NOD2 associated with susceptibility to Crohn's disease. Nature 411(6837):603–606
15. Hampe J, Cuthbert A, Croucher PJ, Mirza MM, Mascheretti S, Fisher S, Frenzel H, King K, Hasselmeyer A, MacPherson AJ, Bridger S, van Deventer S, Forbes A, Nikolaus S, Lennard-Jones JE, Foelsch UR, Krawczak M, Lewis C, Schreiber S, Mathew CG (2001) Association between insertion mutation in NOD2 gene and Crohn's disease in German and British populations. Lancet 357(9272): 1925–1928
16. Hoffman HM, Mueller JL, Broide DH, Wanderer AA, Kolodner RD (2001) Mutation of a new gene encoding a putative pyrin-like protein causes familial cold autoinflammatory syndrome and Muckle-Wells syndrome. Nat Genet 29(3):301–305
17. Wang X, Kuivaniemi H, Bonavita G, Mutkus L, Mau U, Blau E, Inohara N, Nunez G, Tromp G, Williams CJ (2002) CARD15 mutations in familial granulomatosis syndromes: a study of the original Blau syndrome kindred and other families with large-vessel arteritis and cranial neuropathy. Arthritis Rheum 46(11):3041–3045
18. Aganna E, Martinon F, Hawkins PN, Ross JB, Swan DC, Booth DR, Lachmann HJ, Bybee A, Gaudet R, Woo P, Feighery C, Cotter FE, Thome M, Hitman GA, Tschopp J,

McDermott MF (2002) Association of mutations in the NALP3/CIAS1/PYPAF1 gene with a broad phenotype including recurrent fever, cold sensitivity, sensorineural deafness, and AA amyloidosis. Arthritis Rheum 46(9):2445–2452
19. Aksentijevich I, Nowak M, Mallah M, Chae JJ, Watford WT, Hofmann SR, Stein L, Russo R, Goldsmith D, Dent P, Rosenberg HF, Austin F, Remmers EF, Balow JE Jr, Rosenzweig S, Komarow H, Shoham NG, Wood G, Jones J, Mangra N, Carrero H, Adams BS, Moore TL, Schikler K, Hoffman H, Lovell DJ, Lipnick R, Barron K, O'Shea JJ, Kastner DL, Goldbach-Mansky R (2002) De novo CIAS1 mutations, cytokine activation, and evidence for genetic heterogeneity in patients with neonatal-onset multisystem inflammatory disease (NOMID): a new member of the expanding family of pyrin-associated autoinflammatory diseases. Arthritis Rheum 46(12):3340–3348
20. Dode C, Le Du N, Cuisset L, Letourneur F, Berthelot JM, Vaudour G, Meyrier A, Watts RA, Scott DG, Nicholls A, Granel B, Frances C, Garcier F, Edery P, Boulinguez S, Domergues JP, Delpech M, Grateau G (2002) New mutations of CIAS1 that are responsible for Muckle-Wells syndrome and familial cold urticaria: a novel mutation underlies both syndromes. Am J Hum Genet 70(6): 1498–1506
21. Feldmann J, Prieur AM, Quartier P, Berquin P, Certain S, Cortis E, Teillac-Hamel D, Fischer A, de Saint Basile G (2002) Chronic infantile neurological cutaneous and articular syndrome is caused by mutations in CIAS1, a gene highly expressed in polymorphonuclear cells and chondrocytes. Am J Hum Genet 71(1): 198–203
22. Dowds TA, Masumoto J, Chen FF, Ogura Y, Inohara N, Nunez G (2003) Regulation of cryopyrin/Pypaf1 signaling by pyrin, the familial Mediterranean fever gene product. Biochem Biophys Res Commun 302(3): 575–580
23. Goldbach-Mansky R, Dailey NJ, Canna SW, Gelabert A, Jones J, Rubin BI, Kim HJ, Brewer C, Zalewski C, Wiggs E, Hill S, Turner ML, Karp BI, Aksentijevich I, Pucino F, Penzak SR, Haverkamp MH, Stein L, Adams BS, Moore TL, Fuhlbrigge RC, Shaham B, Jarvis JN, O'Neil K, Vehe RK, Beitz LO, Gardner G, Hannan WP, Warren RW, Horn W, Cole JL, Paul SM, Hawkins PN, Pham TH, Snyder C, Wesley RA, Hoffmann SC, Holland SM, Butman JA, Kastner DL (2006) Neonatal-onset multisystem inflammatory disease responsive to interleukin-1beta inhibition. N Engl J Med 355(6):581–592
24. Harton JA, Cressman DE, Chin KC, Der CJ, Ting JP (1999) GTP binding by class II transactivator: role in nuclear import. Science 285(5432):1402–1405
25. Lu C, Wang A, Wang L, Dorsch M, Ocain TD, Xu Y (2005) Nucleotide binding to CARD12 and its role in CARD12-mediated caspase-1 activation. Biochem Biophys Res Commun 331(4):1114–1119. doi:10.1016/j.bbrc.2005.04.027
26. Ye Z, Lich JD, Moore CB, Duncan JA, Williams KL, Ting JP (2008) ATP binding by monarch-1/NLRP12 is critical for its inhibitory function. Mol Cell Biol 28(5): 1841–1850
27. Askari N, Correa RG, Zhai D, Reed JC (2012) Expression, purification, and characterization of recombinant NOD1 (NLRC1): a NLR family member. J Biotechnol 157(1):75–81. doi:10.1016/j.jbiotec.2011.10.007
28. Mo J, Boyle JP, Howard CB, Monie TP, Davis BK, Duncan JA (2012) Pathogen sensing by nucleotide-binding oligomerization domain-containing protein 2 (NOD2) is mediated by direct binding to muramyl dipeptide and ATP. J Biol Chem 287(27):23057–23067. doi:10.1074/jbc.M112.344283

Chapter 13

Blue Native Polyacrylamide Gel Electrophoresis to Monitor Inflammasome Assembly and Composition

Eric M. Kofoed and Russell E. Vance

Abstract

Inflammasomes are large cytosolic multi-protein complexes that form in response to infection and other stimuli, and play an important role in the innate immune response. Traditional methods for assaying inflammasome activation have relied on detection of autoproteolysis of the cysteine protease caspase-1, or proteolytic processing of its substrate, the inflammatory cytokine IL-1β. These measurements report on the final output of inflammasome activation, but do not assess inflammasome assembly and composition. Here we describe the application of blue native gel electrophoresis as a tool for investigating stimulus-dependent inflammasome assembly and for identification of individual components within the multi-protein complex.

Key words Inflammasome, Blue Native PAGE, 2D Native/SDS-PAGE, NAIP, NLRC4, Flagellin

1 Introduction

A key feature of innate responses of both plants and animals is the ability to discriminate between harmful and harmless microbes. This is partly achieved by specific recognition of molecular features that are conserved among classes of potentially pathogenic organisms, but are absent from the host. Both plants and animals use Nucleotide Binding Domain and Leucine-Rich Repeat-containing (NBD-LRR, or NLR) proteins to control infection by pathogenic organisms [1]. In mammals, inflammasomes are composed of sensor NLRs, an adaptor protein called ASC, and the caspase-1 protease that processes inflammatory cytokines and initiates pyroptotic cell death [2, 3]. Formation of this multi-protein signaling platform is a pivotal event during the innate immune response to infection. The specificity, subunit composition, assembly, stoichiometry, and regulation of inflammasomes remain exciting topics of current research, but it has been technically challenging to develop experimental systems with which to address the biochemical properties of inflammasomes.

Christine M. De Nardo and Eicke Latz (eds.), *The Inflammasome: Methods and Protocols*, Methods in Molecular Biology, vol. 1040, DOI 10.1007/978-1-62703-523-1_13, © Springer Science+Business Media New York 2013

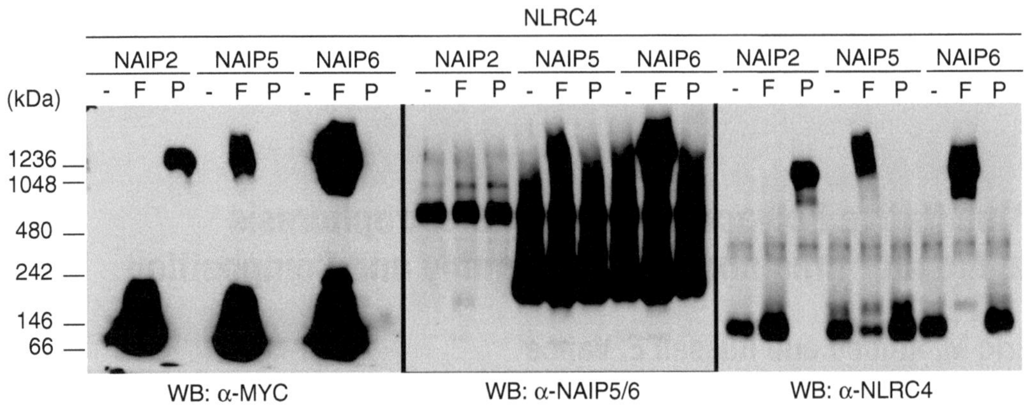

Fig. 1 Blue Native Polyacrylamide Gel Electrophoresis (BN-PAGE) analysis of NAIP-NLRC4 inflammasome assembly and composition. HEK293T cells expressing NLRC4 and NAIP2, NAIP5, or NAIP6, alone or in combination with the bacterial molecules flagellin (6X-MYC-FlaA; F) or PrgJ (6X-MYC-PrgJ; P), were harvested and protein was separated by BN-PAGE. Western blotting for MYC (α-MYC; 9E10™), NAIP5/6, and NLRC4 was performed on a single membrane that was stripped between probes [8]. The blots demonstrate that flagellin induces formation of high molecular weight NAIP5/NLRC4 or NAIP6/NLRC4 inflammasomes, whereas PrgJ induces formation of a high molecular weight NAIP2/NLRC4 inflammasome

Blue native polyacrylamide gel electrophoresis (BN-PAGE) is a method for separating multi-protein complexes from total cell homogenates based on non-denaturing polyacrylamide gel electrophoresis [4, 5]. The BN-PAGE method was pioneered to isolate functional native complexes of the mitochondrial electron transport chain [4], but has found numerous other applications due its elegance and power to interrogate the composition of native protein complexes [6]. The anionic dye Coomassie blue G-250 is mixed with cleared cell lysates to confer a net negative charge to proteins by binding aromatic residues and hydrophobic surfaces, while preserving native protein conformation and detergent-sensitive interactions [4, 5, 7]. Creating protein surfaces bearing an overall negative charge prevents aggregation of hydrophobic proteins, solubilizes membrane proteins, and causes a charge-shift for uniform migration of even basic proteins to the anode [5]. Ultimately, complex mixtures of proteins can be separated and characterized according to molecular size on polyacrylamide gradient gels.

We recently reported that BN-PAGE permits the assessment of NAIP/NLRC4 inflammasome assembly in resting versus stimulated cells (*see* Fig. 1) [8]. The composition of the multi-protein inflammasome can be discerned by subjecting cell lysates separated by BN-PAGE to a second dimension SDS-PAGE and Western blotting for known components (*see* Fig. 2) [8]. Detailed protocols that describe the application of this technique for the discovery of interacting proteins using mass spectrometry can be found elsewhere [6, 9]. The guidelines described below utilize the NativePAGE™ system by Invitrogen that is based on methodology

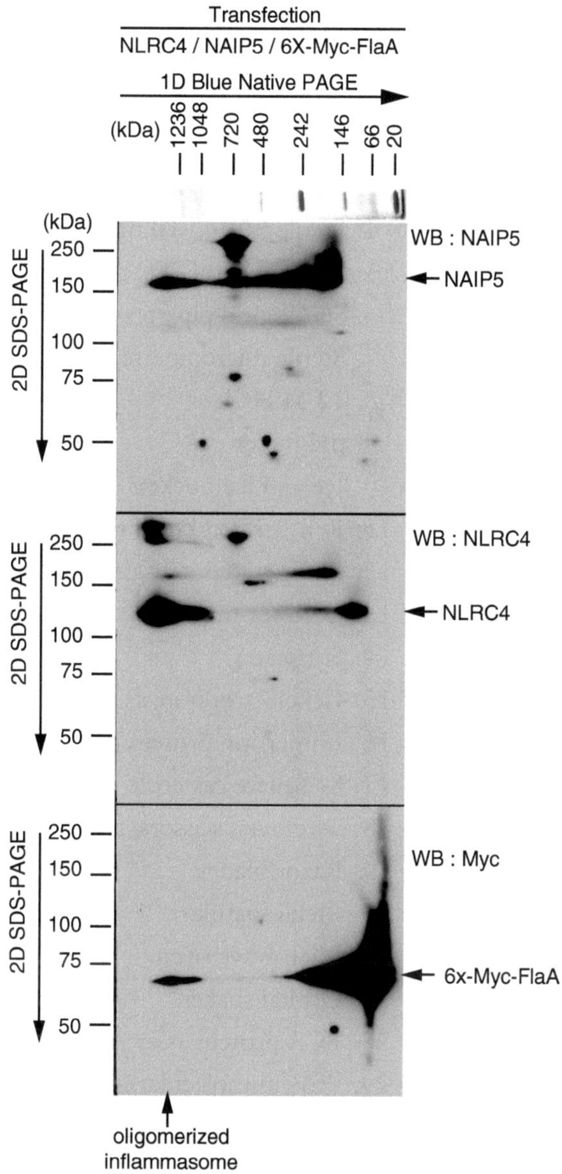

Fig. 2 2D BN-PAGE/SDS-PAGE analysis of NAIP-NLRC4 inflammasome assembly and composition. HEK293T cells expressing NLRC4, NAIP5, and 6X-MYC-FlaA were processed by BN-PAGE after which the lane was cut out and run in a second dimension of SDS-PAGE. Western blotting for NAIP5, NLRC4, and MYC (α-MYC; 9E10™) was performed on a single membrane that was stripped between probes [8]

originally published by Schagger and von Jagow [4, 7]. The Invitrogen pre-cast Novex® gel system eliminates variability in gradient gel preparation, but the methods described below can easily be applied to *in-house* native PAGE systems described in detail elsewhere [4–7, 10, 11].

2 Materials

2.1 General

1. 70 % Ethanol.
2. Spray bottle.
3. Sterile serological pipettes.
4. Sterile 50 and 10 mL conical Tissue Culture (TC) tubes.
5. Sterile 1.5 mL TC tubes.
6. Serological pipettes and micropipettes.
7. Sterile micropipette tips.
8. 12 M HCl.
9. pH meter.
10. Ice and ice buckets.
11. Refrigerated benchtop centrifuge.
12. Refrigerated swinging bucket tissue culture centrifuge.
13. Cold room (4 °C).
14. Stir plates.
15. Gel electrophoresis system.
16. Immersion protein transfer apparatus.
17. 8″-square casserole dish.
18. Tweezers, scissors, and forceps.
19. Razor blades.
20. Metal spatula.
21. Microwave oven.
22. Orbital shaker.
23. BCA protein assay reagent.
24. Protein molecular weight standards.

2.2 Tissue Culture and Transfection

1. Laminar flow hood (HEPA-filtered and UV-sterilized).
2. Cell-culture incubator maintained at 37 °C with 5 %CO_2.
3. Trypan blue solution.
4. Hemocytometer.
5. Tissue culture treated sterile 10 cm petri dishes and 6-well plates.
6. HEK293T cells (ATCC).
7. Complete medium: DMEM, 10 % FBS, 100 U/mL penicillin, 100 μg/mL streptomycin, 2 mM l-glutamine.
8. Liposomal transfection reagent.
9. 1 × 0.05 % Trypsin/EDTA solution, TC grade.

2.3 Sample Preparation

1. Dithiothreitol (DTT).
2. Phosphate buffered saline, TC grade sterile.
3. 5× digitonin stock solution: 5 % digitonin in ddH_2O. Store at −20 °C (*see* **Note 1**).
4. 100× Phenylmethanesulfonyl fluoride (PMSF) stock solution: 100 mM PMSF in 2-propanol. Store at −20 °C (*see* **Note 2**).
5. 100× activated Sodium orthovanadate (Na_3VO_4) stock solution: 200 mM Na_3VO_4 in ddH_2O, pH 10. Store at −20 °C (*see* **Note 3**).
6. 40× Sodium fluoride (NaF) stock solution: 1 M NaF in ddH_2O. Store at −20 °C.
7. 4× Native PAGE base buffer: 200 mM Bis-Tris, pH 7.2, 200 mM NaCl, 40 % glycerol, 0.0004 % Ponceau S. Store at 4 °C.
8. 1× Native PAGE lysis buffer: 50 mM Bis-Tris, pH 7.2, 50 mM NaCl, 10 % glycerol, 0.0001 % Ponceau S, 1 % digitonin, 2 mM Na_3VO_4, 1 mM PMSF, 25 mM NaF, 1× Complete™ EDTA-free protease inhibitor cocktail (Roche). Make 1× native PAGE lysis buffer just prior to cell collection by adding detergent and protease inhibitors fresh from frozen stock solutions (*see* **Note 4**).
9. Coomassie brilliant blue G-250 sample additive: 2.5 % Coomassie G-250 in ddH_2O. Store at −20 °C.

2.4 One-Dimensional Blue Native PAGE

1. NativeMark™ Unstained Protein Standard (Invitrogen).
2. NativePAGE™ Novex® 3–12 % Bis-Tris gels, 1.0 mm, 10-well (Invitrogen).
3. Power supply.
4. 1× Native PAGE running buffer: 50 mM Bis-Tris, 50 mM Tricine, pH 6.8 (*see* **Note 5**).
5. 20× Native PAGE cathode buffer additive stock: 0.4 % Coomassie G-250 in 1× Native PAGE running buffer. Sterile filter and store at 4 °C (*see* **Note 6**).

2.5 Second-Dimension (2D) SDS-PAGE

1. NuPAGE 4–12 % Bis-Tris Gels, 1.0 mm × 2D well (Invitrogen).

2.6 Western Blotting

1. Anti-Flag monoclonal antibody (M2).
2. Polyvinylidene difluoride (PVDF) membrane.
3. c-Myc (9E10) monoclonal antibody.
4. HRP-conjugated secondary antibody.
5. ECL solution.
6. 3 M chromatography paper (Whatman).

7. X-ray film.
8. Developer and dark room.
9. Square petri dishes.
10. Bromophenol blue.
11. 100 % Methanol.
12. 1× Laemmli SDS-PAGE gel loading buffer: 62.5 mM Tris–HCl, pH 6.8, 2 % SDS, 10 % glycerol, 0.1 % bromophenol blue, 200 mM DTT (*see* **Note 7**).
13. 1× Western blotting transfer buffer: 48 mM Tris-base, 39 mM glycine, 0.037 % SDS, 20 % methanol (*see* **Note 8**).
14. 1× Tris-buffered saline with Tween-20 (TBST): 20 mM Tris-base, pH 7.6, 137 mM NaCl, 0.2 % Tween-20 (*see* **Note 9**).
15. 1× MOPS SDS-PAGE running buffer: 50 mM MOPS, 50 mM Tris-base, 3.5 mM SDS, 1 mM EDTA (*see* **Note 10**).
16. Blocking solution: TBST + 5 % nonfat dry milk.

2.7 Stripping Western Blots

1. 1× Mild stripping buffer: 200 mM glycine, pH 2.2, 0.1 % SDS, 1 % Tween-20.

3 Methods

3.1 Cell Culture, Transfection, and Lysis

1. Culture HEK 293T cells in complete medium and seed into 6-well tissue-culture plates at 8×10^5 cells per well.
2. The next day, transfect cells using Lipofectamine2000™ reagent at a ratio of 2 μg plasmid DNA to 5 μL reagent according to the manufacturer's instructions (*see* **Notes 11** and **12**).
3. Transfect plasmids encoding known inflammasome components alone, or in combination, and equalize the total amount of transfected DNA with empty vector (*see* **Note 13**).
4. Collect cells 48 h after transfection for use in various assays. Aspirate culture medium, and wash cells with 2 mL cold PBS. Aspirate PBS wash and add 300 μL 0.05 % trypsin/EDTA per well, and place cells at 37 °C for 5 min (*see* **Note 14**).
5. Neutralize trypsin by adding 700 μL complete medium, transfer cells to 1.5 mL tubes and pellet at 4 °C, 400 × *g*, for 5 min.
6. Aspirate culture supernatant and wash cells 3× with 1 mL cold PBS to remove traces of trypsin (*see* **Note 15**).
7. Resuspend cell pellet in 200 μL 1× native PAGE lysis buffer that contains detergent and fresh inhibitors (*see* **Note 16**).
8. Triturate 10× per sample using a 200 μL pipette, 3× per sample during a 30 min period at 4 °C.
9. Centrifuge samples at 16,000 × *g* for 30 min at 4 °C in a benchtop centrifuge.

10. Transfer soluble protein lysates to new cold 1.5 mL tubes careful not to disturb the cellular debris pellet.
11. Perform BCA protein quantification assay on samples to determine total protein concentration and adjust the samples to equalize the total protein between samples (*see* **Note 17**).

3.2 Blue Native Gel Electrophoresis

1. Prepare samples for Native PAGE by mixing 2.5 μL of Coomassie G-250 sample additive with 22.5 μL of cell lysate equalized for total protein, and maintain samples on ice at all times (*see* **Note 18**).
2. In the cold room (4 °C), remove the white tape from the foot of NativePAGE™ Novex® Bis-Tris minigels and gently remove the comb under cold running water to prevent damage to the lane dividers. Flush wells with 1× native PAGE running buffer prior to loading samples.
3. Assemble the PAGE apparatus in the cold room on a stir-plate with a stir-bar, locking both NativePAGE™ Novex® Bis-Tris minigels into place so that the open wells are contiguous with the cathode (inner chamber) buffer (*see* **Notes 19** and **20**).
4. Fill the inner and outer chambers with ice-cold 1× native PAGE running buffer to cover the sample wells.
5. Load 5 μL of unstained NativeMark™ protein standard and 25 μL of prepared samples into each well (*see* **Note 21**).
6. After all samples are loaded, remove 10 mL cathode running buffer with a serological pipette, and *carefully* add 10 mL of 20× Cathode Buffer Additive Stock to generate a "dark-blue" cathode buffer (0.02 % G-250). Mix gently by trituration to avoid sample disturbance (*see* **Note 22**).
7. Run at 150 V constant for 60 min. Current starts at ~15 mA per gel.
8. Stop the run and decant the "dark-blue" cathode buffer. Fill the cathode chamber with 1× NativePAGE™ Running Buffer, and add 1 mL of 20× Cathode Buffer Additive to generate a "light-blue" cathode buffer (0.002 % G-250) (*see* **Note 23**).
9. Run at 250 V constant for 120 min. Current starts at ~20 mA per gel (*see* **Note 24**).
10. Skip to **step 1** in Subheading 3.4 if Western blotting one-dimensional native PAGE.

3.3 Blue Native/SDS Two-Dimensional PAGE

1. Place a glass casserole dish (8″ × 8″) in the fume hood and fill with ~500 mL of 1× MOPS SDS-PAGE Running Buffer. Gather two pairs of tweezers, plastic ruler, marker, and as many NuPAGE™ 4–12 % Novex® Bis-Tris minigels (1.0 mm × 2D well) as samples you intend to run in the second dimension (*see* **Note 25**).

2. Pry open the casting plates from the NativePAGE™ gel (resume protocol from **step 9** in Subheading 3.2) using a metal spatula, remove the foot of the gel and the well dividers.
3. Slice the gel from the top (wells) to the bottom (foot) using a razor blade, cutting down both sides of the lane containing the sample of interest.
4. Trim the gel slices to fit the specifications of your 2D well, making sure you cut the gel such that the slice (or lane) will be ~3 mm shorter than the length of the 2D well (*see* **Note 26**).
5. Place each gel slice into a pre-labeled individual square petri dish containing 5 mL of 1× Laemmli SDS-PAGE gel loading buffer and shake at room temperature (RT) for 10 min (*see* **Note 27**).
6. Move the petri dish containing the gel slice to a microwave and boil on high setting for ~20 s.
7. Continue to shake the gel slice at RT for another 10 min.
8. Remove the comb of a 2D gel in the 1× MOPS SDS-PAGE running buffer ready from **step 1** in Subheading 3.3.
9. Move one gel slice into the glass casserole dish and immerse in MOPS buffer (*see* **Note 28**).
10. Orient the gel slice to the 2D gel so that the high MW complexes are closest to the protein standard lane (2D gels have a single very wide well, and one small well for the SDS-PAGE protein standard).
11. Insert the 12 % acrylamide end of the gel slice at a 45° angle to the 2D well, gently pushing the gel slice down and towards the side opposite the protein standard. Position the gel slice flush with the corner, then gently massage the 3 % side of the gel slice into the well (*see* **Note 29**).
12. Run the NuPAGE™ Bis-Tris 4–12 % 2D SDS-PAGE gel at 150 V at RT until the dye front nears the bottom.

3.4 Protein Transfer and Western Blotting

1. Fill a glass casserole dish with ~500 mL 1× Western Blotting Transfer Buffer, and submerge the protein transfer apparatus and several sponges. Assemble a transfer sandwich by placing one sponge on the negative plate and two pieces of Whatman 3 M paper cut to the exact dimensions of your gel.
2. Disassemble the PAGE apparatus, remove the plastic casing from the gel using a metal spatula, cut off the gel foot and trim the gel ~2 mm below the well bottom. Cut a small triangle off the bottom corner to orient the samples after transfer to PVDF membrane (*see* **Notes 30** and **31**).
3. Transfer the trimmed gel to the transfer cassette using a razor blade or tweezers and your gloved fingers to pick up gel from the bottom (*see* **Note 32**).

4. Cut the PVDF membrane to match the size of the gel and activate it in 100 % methanol for 1 min. followed by equilibration in 1× Transfer Buffer for 1 min.
5. Carefully layer the activated PVDF onto the gel, and use an elbow-tweezer to smooth out any bubbles that exist between membrane and gel (*see* **Note 33**).
6. Overlay two pieces of Whatman paper followed by sponges filled flush to the top of the blot module.
7. Perform a wet transfer overnight (~12 h) at 15 V constant (~ 100 mA per transfer module) in the cold room (*see* **Notes 34** and **35**).
8. Open transfer apparatus and separate the membrane from the gel. Cut the corner of the membrane to retain the sample orientation (the PVDF is now blue, but the dye will disappear during subsequent steps).
9. Place PVDF membrane protein-face up in a dish containing 1× TBST to wash ~1 min. Decant wash and replace with blocking solution, and shake for 1 h at RT.
10. Probe membrane for specific proteins using primary antibodies. Epitope-tagged protein expressed in HEK293T cells are detected using primary antibody (e.g., anti-FLAG (M2™) or anti-myc (9E10™)) at 1:5,000 in 10 mL blocking solution, rocking at room temp for ~3 h (*see* **Note 36**).
11. Wash membrane 3× with 10 mL TBST, 10 min each wash.
12. Incubate membrane for 1 h with 10 mL blocking solution containing a 1:5,000 dilution of HRP-conjugated secondary antibody (raised against the species in which the primary antibody was generated).
13. Wash 3× with 10 mL TBST, 10 min each wash.
14. Develop blot using enhanced chemi-luminescence (ECL) according to the reagent manufacturer (*see* **Note 37**).

3.5 Stripping and Re-probing Western Blots

1. Membranes can be stripped and re-probed for different inflammasome components. Alignment of these Western blots reveal whether different proteins are contained in the same complex (*see* Figs. 1 and 2) (For controls *see* **Notes 38** and **39**).
2. Following ECL, rinse membranes in 10 mL 1× TBST, and place into a dish containing 10 mL 1× mild stripping buffer for 10 min at RT with shaking.
3. Discard buffer and replace with fresh 10 mL 1× mild stripping buffer at RT with shaking.
4. Discard buffer and wash 2× in 10 mL 1× PBS for 10 min each wash.

5. Discard buffer and wash 2× in 10 mL 1× TBST for 10 min each wash.
6. Resume Western blotting procedure at the blocking step (*see* **step 9** in Subheading 3.4 above).

4 Notes

1. If any precipitation of detergent stock solution occurs, briefly warm in 37 °C water bath (or 95 °C heat block) to solubilize.
2. Warm in 37 °C water bath to solubilize.
3. Prepare 200 mM solution of sodium orthovanadate in fume hood. Dissolve Na_3VO_4 in ddH_2O and adjust to pH 10. The solution will be yellow in color. Boil until the solution becomes colorless (~10 min) then cool to RT. If solution is yellow after cooling, adjust to pH 10 and repeat boiling step. Bring to volume, aliquot, and store at −20 °C. Note: DTT inactivates this tyrosine-phosphatase inhibitor.
4. Add protease inhibitors fresh to lysis buffer from frozen stock solutions, as many have short half lives in working solution at neutral pH and room temperature.
5. Prepare a 20× stock and store at RT.
6. Used as a 20× stock this additive makes the Native PAGE Running Buffer into the "dark cathode buffer" (0.02 % (w/v) G-250); used as a 200× stock this makes "light cathode buffer" (0.002 % (w/v) G-250).
7. Make up a 6× stock of Laemmli loading buffer, store aliquots at −20 °C. Add DTT just before use from 1 M DTT stock.
8. Make up 10× stock of Western blotting transfer buffer (without methanol) and store at RT. Add methanol (to 20 %) at time of final dilution.
9. Make up 20× stock of TBST and store at RT.
10. Make up 20× stock of MOPS SDS-PAGE running buffer and store at RT. Do not adjust pH.
11. The concentration of transfected NLR plasmid cDNAs required to observe ligand-inducible inflammasome assembly will vary according to the expression level and the NLR. However, successful flagellin-dependent assembly of the NAIP5-NLRC4 inflammasome using the murine stem cell virus (MSCV) retroviral expression vector is achieved by delivering 400 ng MSCV2.2-NAIP5, 400 ng MSCV2.2-NLRC4, 400 ng MSCV2.2-flagellin, and 800 ng MSCV2.2 empty vector into ~1×10^6 HEK293T cells.
12. Using HEK293T cells, there is no need to change the media during the experiment, because Lipofectamine2000™ used at

this concentration is not toxic to these cells. However, when transfecting other cell lines for the first time it is important to test the toxicity of the reagent empirically.

13. If monitoring inflammasome assembly it is important to omit mCaspase-1 cDNA to prevent pyroptotic cell death. Alternatively, if the aim is to monitor pyroptosis, then co-transfection of mCaspase-1 is required. Transfecting 400 ng of each inflammasome component expressed from the MSCV LTR is a good starting point for monitoring ligand-dependent assembly, but use of other promoters (e.g., CMV) and different NLRs will probably require titrating expression levels empirically, to avoid ligand-independent inflammasome assembly.
14. If monitoring transmembrane proteins or complexes do not use trypsin. Instead detach adherent cells by incubation with 0.5 mM EDTA in PBS for 5 min at 4 °C.
15. During washes aspirate supernatant down to ~200 μL per well above the cell pellet to avoid accidental aspiration. Use a 200 μL pipette tip to carefully remove the final wash. Keep samples on ice at all times during cell lysis and use a refrigerated (4 °C) centrifuge. Chill all buffers and run native samples the day they are collected whenever possible to avoid freeze–thaw of samples.
16. Detergents should be chosen empirically. Most commonly used detergents and respective concentration ranges (w/v) in native lysis buffers are digitonin (0.5–1.0 %), Triton X-100 (0.1–0.5 %), Brij 96 (0.1–0.5 %), dodecylmaltoside (0.1–0.5 %) (*see* Fig. 3) [11].
17. Reducing agents are not compatible with the BCA reagent.
18. Samples prepared for native PAGE require addition of Coomassie G-250 to 1/4th the detergent concentration. If digitonin concentration is 1 %, the final dye concentration needed is 0.25 % G-250. Add an appropriate volume of Coomassie G-250 sample additive to each sample, gently vortex, and quick-spin to collect the entire volume for loading.
19. Excellent step-by-step procedures for preparation of *in-house* gradient native gels can be found in Shagger and von Jagow [4, 7] and Wittig et al. [5]. However, gels and buffers are not interchangeable.
20. Use a stir-bar when performing native PAGE, because different concentrations of salt along the apparatus bottom can cause aberrant protein migration.
21. Invitrogen NativePAGE™ Novex® Bis-Tris minigels are sold as precast 3–12 % and 4–16 % gradients, in 10-well format and

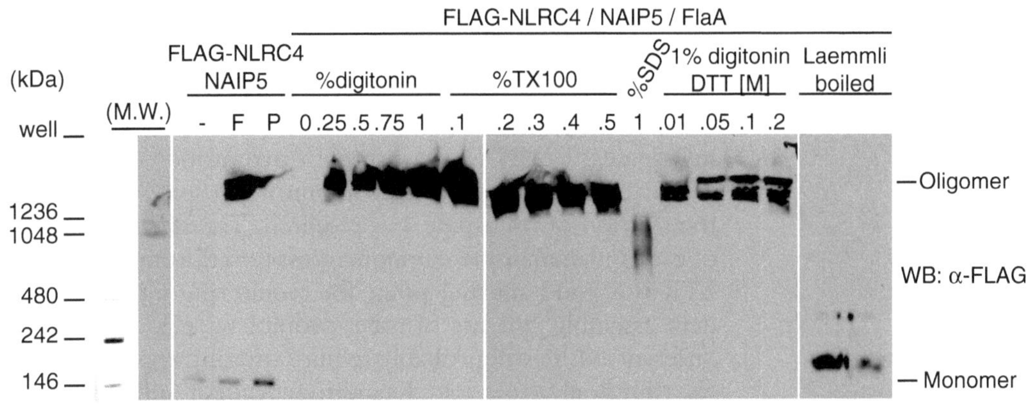

Fig. 3 Influence of detergent and reducing agent concentration on the ability to measure NLRC4 inflammasome formation by Blue Native PAGE. HEK293T cells expressing FLAG-NLRC4, NAIP5, and either FlaA (F), or PrgJ (P) were harvested in lysis buffers with varying concentration of digitonin (0–1 %), TX100 (0.1–0.5 %), SDS (1 %), or 1 % digitonin with DTT (0.01–0.2 M) and protein was separated by Blue Native PAGE and Western blotted for FLAG-NLRC4 using the M2™ (Sigma) antibody. NAIP5-NLRC4 Inflammasome assembly occurs in response to FlaA, but not PrgJ [8]. All concentrations of digitonin and TX100 efficiently extract inflammasomes in their native state, whereas SDS (1 %) does not. Reducing agent (DTT) concentration above 50 mM causes a minor shift of complexes around 1,200–1,400 kDa, but does not significantly alter the stability of the inflammasome. Boiling samples in 1× Laemmli SDS-PAGE sample buffer for 5 min completely denatures the oligomeric complex

1.0 mm thickness. The maximum well volume is 25 μL. Outlining well bottoms with a marker aids in visualization.

22. If the 20× native PAGE cathode buffer additive stock is added to the cathode chamber before loading the samples it is impossible to visualize individual wells for sample loading. We use the XCell Surelock™ electrophoresis system, which has a cathode chamber volume of 200 mL.

23. For Western blotting applications, changing the cathode buffer from "dark," to "light" after the gel has run for 1 h at 150 V flushes excess dye from the gel so the PVDF membrane will not be blocked by excessive dye when it comes time to transfer protein. Transfer conditions that benefit larger proteins and complexes (i.e., low voltage overnight transfer) may allow smaller proteins to pass through the membrane. This can manifest if the PVDF is clogged with excess G-250 dye.

24. The dye front takes approximately 3 h to reach the foot of the gel. NativeMark™ protein standards are unstained, but B-phycoerythrin (252 kDa) appears as a red band visible under normal fluorescent light.

25. Performing a second dimension of SDS-PAGE after a first dimension of blue native PAGE will require a separate 2D

SDS-PAGE gel for every sample well loaded in the first dimension. Plan accordingly.

26. The 2D well of a NuPAGE™ minigel is 6 cm long. Measure (well to foot) using a ruler and cut the gel so that each lane will yield a ~5.7 cm long gel slice. Samples that are to be compared must be run on the same native gel to ensure the gel slices represent comparable migration distances that you can align when performing downstream applications. Therefore, it is important to decide whether 5.7 cm from the top or bottom of the native gel will be represented in the second dimension.
27. Laemmli SDS-PAGE sample buffer contains β-mercaptoethanol or DTT, which are volatile and irritating to the olfactory system. If possible, reduce the amount of time outside the hood, or place orbital shaker in the hood.
28. Insertion of the gel slice into the 2D well takes place submerged in buffer to prevent air bubbles from getting trapped between the gels. Contact between the native gel slice and SDS-PAGE gel must be flush for even passage.
29. Tweezers can aid this process. Inserting tweezers into the 1 mm 2D well, and gently rotate them to open the width of the 2D well just enough to get the gel slice inserted. If the gel slice doesn't fit into the 2D SDS-PAGE well, it must be removed and trimmed.
30. It may be useful to keep the wells intact during preliminary experiments to check if protein complexes are too large to be separated or suffering from aggregation. Technical problems can usually be overcome by using different detergent, ultracentrifugation, or dialysis prior to Native PAGE. For example, 4–16 % gels cannot effectively separate a ~1,200 kDa complex, whereas the 3–12 % gels do permit entry and separation of a protein complex this size.
31. Standards can be visualized on PVDF membrane by colloidal blue, or the marker lane can be excised from the gel prior to transfer and stained overnight with Coomassie R-250, destained, and gel dried. The 1,246 and 1,038 kDa markers can be detected through background reaction with HRP-conjugated secondary antibody upon longer exposures (*see* Fig. 3).
32. Prevent tearing the gel by lifting from the 12 % foot, and not the 3 % wells.
33. An alternative to using an elbow-tweezer to remove bubbles between the membrane and gel is to gently roll out bubbles using a pipette. Extreme caution must be used to avoid deforming the 3 % end of gradient gels. Anchor the membrane at the 12 % side with gloved fingers when expelling bubbles.

34. Longer transfer at low current is preferable for large protein complexes, but 35 V constant (current starting at ~250 mA per transfer module) for 90 min at RT also works well.
35. If the antibody to be used in subsequent Western blotting was raised against a masked (or internal) epitope of a native protein, place gel in a dish containing a solution of 10 % SDS for 5 min, gently rocking, prior to transfer. This step completely denatures protein complexes prior to transfer, revealing all available epitopes, but may also cause protein bands to smear or appear fuzzy.
36. Alternatively, probe with primary antibody overnight at 4 °C with shaking.
37. Mix 2 mL reagent A (hydrogen peroxide), and 2 mL reagent B (luminol) of ECL solution for 1 min, soak membrane protein-face up for 1 min, then place in sheet protector taped to the inside of a X-ray film cassette, and develop X-ray film in the dark room for 30 s, 2 min, and 5 min.
38. Dissociation of multi-protein complexes (adding 1 % SDS, or boiling at 95 °C for 5 min in 1× Laemmli SDS-PAGE sample buffer) is a good control that can be run alongside the native sample in the first dimension blue native PAGE (*see* Fig. 3). Skip a lane between these samples so the detergent doesn't affect neighboring lanes.
39. It is conceivable that the protein components are coincidentally in two separate complexes that happen to be of the same size. Using an antibody to shift the complex, affinity copurification, yeast-two-hybrid, and in vitro binding assays using purified protein are complementary techniques of demonstrating specific association.

Acknowledgments

E.K. would like to thank Dr. Maryanne McClellan who taught Western blotting in an era of *in-house* ECL solution when X-ray film was developed by hand, and Dr. Elizabeth Vierling and Dr. Gary Lee who taught E.K. native PAGE a decade ago. Genentech kindly provided the α-NLRC4 antibody, and Bill Dietrich contributed the α-NAIP5 antibody.

References

1. Bonardi V, Cherkis K, Nishimura MT, Dangl JL (2012) A new eye on NLR proteins: focused on clarity or diffused by complexity? Curr Opin Immunol 24(1):41–50. doi: S0952-7915(11)00181-6 [pii] 10.1016/j.coi.2011.12.006
2. Bergsbaken T, Fink SL, Cookson BT (2009) Pyroptosis: host cell death and inflammation.

Nat Rev Microbiol 7(2):99–109. doi:nrmicro 2070 [pii] 10.1038/nrmicro2070
3. Martinon F, Burns K, Tschopp J (2002) The inflammasome: a molecular platform triggering activation of inflammatory caspases and processing of proIL-beta. Mol Cell 10(2): 417–426. doi:S1097276502005993 [pii]
4. Schagger H, von Jagow G (1991) Blue native electrophoresis for isolation of membrane protein complexes in enzymatically active form. Anal Biochem 199(2):223–231
5. Wittig I, Braun HP, Schagger H (2006) Blue native PAGE. Nat Protoc 1(1):418–428. doi:nprot.2006.62 [pii] 10.1038/nprot. 2006.62
6. Lasserre JP, Menard A (2012) Two-dimensional Blue Native/SDS gel electrophoresis of multiprotein complexes. Methods Mol Biol 869:317–337. doi:10.1007/978-1-61779-821-4_27
7. Schagger H, Cramer WA, von Jagow G (1994) Analysis of molecular masses and oligomeric states of protein complexes by blue native electrophoresis and isolation of membrane protein complexes by two-dimensional native electrophoresis. Anal Biochem 217(2): 220–230. doi:S0003269784711122 [pii]
8. Kofoed EM, Vance RE (2011) Innate immune recognition of bacterial ligands by NAIPs determines inflammasome specificity. Nature 477(7366):592–595. doi:[pii] 10.1038/nature10394
9. Susnea I, Bernevic B, Wicke M, Ma L, Liu S, Schellander K, Przybylski M (2012) Application of MALDI-TOF-mass spectrometry to proteome analysis using stain-free gel electrophoresis. Top Curr Chem. doi:10.1007/128_2012_321
10. Swamy M, Kulathu Y, Ernst S, Reth M, Schamel WW (2006) Two dimensional Blue Native-/SDS-PAGE analysis of SLP family adaptor protein complexes. Immunol Lett 104(1–2): 131–137. doi:S0165-2478(05)00350-0 [pii] 10.1016/j.imlet.2005.11.004
11. Swamy M, Siegers GM, Minguet S, Wollscheid B, Schamel WW (2006) Blue native polyacrylamide gel electrophoresis (BN-PAGE) for the identification and analysis of multiprotein complexes. Sci STKE 2006(345):pl4. doi:stke.3452006pl4 [pii] 10.1126/stke. 3452006pl4

Chapter 14

Analysis of Microbiota Alterations in Inflammasome-Deficient Mice

Eran Elinav, Christoph A. Thaiss, and Richard A. Flavell

Abstract

Inflammasomes have emerged as central regulators of intestinal infection, immunity, and inflammation. Inflammasome activity mediates intestinal epithelial integrity, antimicrobial responses, and initiates inflammation through generation of the cytokines interleukin (IL-)1 and IL-18. Recent studies have identified an additional layer of inflammasome function in the intestine, namely, the control of intestinal microflora composition. Inflammasome-deficient mice show an aberrant microbial community which is dominantly transmissible to healthy mice. This dysbiosis in inflammasome-deficient mice has a profound impact on their physiology and pathophysiology, both locally in the intestine and systemically. Therefore, it is essential to consider the influence of the composition of microbial communities on experiments performed with inflammasome-deficient and other innate molecule-deficient mice, and to conduct experiments to control for potential dominant effects of the microflora on host responses. In this chapter, we provide experimental procedures to monitor inflammasome-mediated modifications of the intestinal microflora composition in mice and to test the resultant functional consequences of these changes in microbial communities and their transmission to cohoused mice.

Key words Inflammasome, Microflora, Dysbiosis, Intestine, Cohousing, 16S RNA sequencing

1 Introduction

The formation of a multi-molecular complex consisting of an upstream NLR protein, the adaptor molecule ASC, and caspase-1—named inflammasome—was first described by Juerg Tschopp and colleagues 10 years ago [1]. Since then, many biochemical and cellular features of inflammasomes have been revealed: they have been associated in several physiological and pathophysiological processes, and their mechanisms of action and triggering stimuli are under intense investigation [2]. However, we are just beginning to understand the tissue and cell type-specific roles that inflammasome complexes play in vivo. As sensors of both microbe-associated

Eran Elinav and Christoph A. Thaiss contributed equally to this work.

Christine M. De Nardo and Eicke Latz (eds.), *The Inflammasome: Methods and Protocols*, Methods in Molecular Biology, vol. 1040, DOI 10.1007/978-1-62703-523-1_14, © Springer Science+Business Media New York 2013

molecular patterns and endogenous signals of damage [3], one would predict that inflammasomes exert critical functions in organ systems which are potential entry sites for infectious pathogens, and are exposed to both microbial colonization and mechanical stress. Indeed, recent studies have identified important functions for inflammasomes at mucosal surfaces, ranging from pivotal functions in tissue homeostasis to host defense to infection [4]. Importantly, inflammasomes are involved in regulating the composition of intestinal microbial communities, and inflammasome-deficient mice feature an aberrant microbiota [5], with several specific bacterial taxa being overrepresented as compared to the "regular" microflora of a wild-type mouse (although the latter may vary significantly among animal facilities, housing conditions and diet). This aberrant microbiota triggers an enhanced inflammatory reaction in the intestine and predisposes the host to inflammatory bowel disease [5]. Moreover, inflammasome-deficient mice feature compromised intestinal barrier function, and microbial components are more prone to access the circulation, thereby also affecting distal organ systems, most prominently the liver. There, the presence of microbial products causes a basal inflammatory state, which predisposes the host to manifestations of the metabolic syndrome, such as insulin resistance and hepatic steatosis [6].

Notably, this aberrant microbial community in inflammasome-deficient mice is dominantly transferred to wild-type mice housed in the same cage, and causes the same disease manifestations in inflammasome-competent recipient mice [5]. For instance, when cohoused with inflammasome-deficient mice, wild-type mice become more susceptible to the development of nonalcoholic fatty liver disease, and show a similarly enhanced penetration of microbial products into the circulation [6]. This shows that inflammasome deficiency may not be the direct cause of disease susceptibility, but rather initiates intestinal dysbiosis, which then predisposes the host to disease. Therefore, it is extremely important to consider the impact of dysbiosis on any experimental outcome when studying the effects of inflammasome deficiencies in vivo.

Inflammasomes are not the only sensor molecules of the innate immune system which have been recognized for their role in shaping the composition of the intestinal microbiota. For instance, aberrant microbial compositions and properties have been reported for mice lacking TLR5 and MyD88 [7], indicating that the sensing of microbes, which are in contact with intestinal epithelial cells or have penetrated into the lamina propria, is crucial to maintaining a mutualistic relationship between the host and its microbial environment. Furthermore, the loss of this sensing ability results in outgrowth and abnormal behavior of bacteria that would be defined as commensals in healthy hosts [8].

Given the increasing recognition of the importance of the intestinal microbiota for both local and systemic host physiology in health and disease, it is essential that certain experimental procedures be followed to analyze the contribution of the intestinal microbial composition to disease phenotypes of inflammasome-deficient mice. We describe here methods which are useful to monitor the intestinal microbial composition in inflammasome-deficient mice, and to test the functional impact of dysbiosis on physiology and pathophysiology. We focus on three major procedures: (1) transfer of the microflora by cohousing, cross-fostering, and oral gavage, (2) sampling and processing of fecal material, and (3) DNA sequencing and analysis methods.

2 Materials

2.1 Mice

1. C57BL/6.
2. $ASC^{-/-}$, Caspase-$1^{-/-}$.
3. IL-$18^{-/-}$, IL-$1\beta^{-/-}$.
4. $NLRP3^{-/-}$, $NLRP6^{-/-}$.

2.2 Consumables

1. 2 mL screw-cap tube (Axygen).
2. 1.5 mL tubes.
3. 0.1 mm glass beads, density 2.5 g/cc (BioSpec).
4. Phase-lock gel (PLG) tube (5′).
5. Plastic pestle.
6. 40 μm cell strainer (BD).
7. Gavage needle.
8. Sterile forceps.

2.3 Instruments

1. Mini-bead beater (BioSpec).
2. qPCR cycler.
3. Water bath or heat block at 50 °C.

2.4 Reagents

1. TNE Buffer: 200 mM Tris–HCl, pH 8.0, 200 mM NaCl, 20 mM EDTA.
2. 20 % SDS solution.
3. Phenol–chloroform–isoamyl alcohol solution (25:24:1), pH 7.9.
4. Ice-cold isopropanol.
5. 3 M NaOAc, pH 5.5.
6. 100 % ethanol.
7. Phosphate buffered saline (PBS).
8. TE buffer: 10 mM Tris–HCl, 1 mM EDTA, pH 8.0.
9. Universal qPCR mix (Kapa Biosystems).
10. Liquid nitrogen.

11. Quant-iT PicoGreen DNA Kit (Invitrogen).
12. Agencourt AMPure XP PCR purification kit (Beckman Coulter).

3 Methods

3.1 Equilibration of the Microflora

3.1.1 Cohousing

The coprophagic behavior of mice makes them an ideal model to study the functional consequences of bacterial transfer to mice housed in the same cage. To allow for full equilibration of the composition of the intestinal microflora, the following protocol has proven effective:

1. After weaning, cohouse inflammasome-deficient mice and wild-type mice in a 1:1 ratio.
2. Maintain cohabitation for at least 4 weeks, to allow for complete transfer of the microbiota (*see* **Note 1**).
3. Perform the experiment of interest while keeping the mice cohoused, as separation may lead to return of the original microbiota composition (*see* **Note 2**).

3.1.2 Cross-Fostering

Shortly after birth, the mucosal surfaces of newborns are colonized by maternally transmitted bacteria [9]. In addition, maternal IgA and antimicrobial peptides are transferred to newborns by breastfeeding, further influencing the establishment of intestinal microbial communities [10]. It is therefore possible to equip wild-type animals with microbial communities from inflammasome-deficient mothers and vice versa, using the following protocol:

1. Set up breeding pairs of wild-type and inflammasome-deficient strains simultaneously, so that litters can be expected on the same day.
2. Within 24 h of birth, swap newborns between both breeding pairs, so that wild-type pups will be cross-fostered by inflammasome-deficient dams and vice versa. Swapping of newborn animals between cages should be carried out with sterile forceps. Avoid touching the newborns.
3. After 3 weeks of cross-fostering, the pups can be weaned and the experiment of interest can be carried out while keeping the genotypes matched.

3.1.3 Transfer of Microbial Communities by Oral Gavage

A third way of imposing a foreign microbial community on recipient mice is to orally gavage a suspension of fecal microbiota.

1. Collect fecal pellets into a tube. Alternatively, when sacrificing the donor animal, collect cecal content.
2. Add 50 μL of PBS to each fecal pellet and homogenize manually with a pestle.

3. Filter the sample through a 40 μm cell strainer.
4. Transfer the sample to recipient mouse by oral gavage (*see* **Note 3**).

The three procedures described so far offer possibilities to equilibrate the microflora between wild-type and inflammasome-deficient mice, and to test whether the intestinal bacterial composition influences their predisposition to a certain phenotype (*see* **Note 4**). Often, however, it is of interest to determine the bacterial taxa which are differentially represented in different microbial communities. In the next section, we describe experimental procedures to harvest and process samples for comprehensive analysis of intestinal bacterial communities. We will then provide an overview of currently available sequencing and analysis methods for the study of the microbiome.

3.2 Analyzing Intestinal Microbial Communities

3.2.1 Sampling and Processing of Fecal Material

1. Add 250 μL of 0.1 mm beads to each 2 mL screw-cap tube. Record the weight.
2. Collect fresh feces from mice into collection tubes and freeze immediately in liquid nitrogen.
3. Transfer the sample from the collection tube to a bead-containing tube. This can be done by placing them end-to-end and flipping the sample into the new tube. Record the weight of the full tube. Calculate net weight of sample.
4. Add 500 μL of TNE buffer.
5. Add 210 μL of 20 % SDS.
6. Add 500 μL of phenol–chloroform–isoamyl alcohol solution. Keep samples on ice.
7. Homogenize the sample for 2 min using a mini-bead beater (*see* **Note 5**).
8. Centrifuge the tubes at 5,000 × *g* for 3 min at 4 °C.
9. In the meantime, spin down PLG columns at 14,000 × *g* for 30 s at room temperature.
10. Transfer the aqueous phase (around 600 μL) to the pre-spun PLG tube.
11. Add 600 μL of phenol–chloroform–isoamyl alcohol solution to the sample and mix by inversion (do not vortex).
12. Centrifuge the tubes at 14,000 × *g* for 5 min at room temperature.
13. Transfer the aqueous phase to a new 1.5 mL tube.
14. Add 60 μL (i.e., 10 % of final vol) of 3 M NaOAc and mix thoroughly by inversion.
15. Add 600 μL of ice-cold isopropanol.

16. Store at −20 °C overnight (*see* **Note 6**).
17. Centrifuge tubes at 14,000 × *g* for 20 min at 4 °C. Decant supernatant.
18. Wash the pellet (which may range from dark brown to off-white, depending on the diet the animal was on) with 500 μL of 100 % ethanol.
19. Centrifuge for 3 min at 4 °C. Decant the supernatant.
20. Gently remove excess ethanol.
21. Suspend pellet in TE buffer using 5 μL per mg fecal pellet material used in isolation.
22. Incubate for 30 min at 50 °C. Vortex every 10 min to dislodge pellet from tube wall.
23. Store at −20 °C until further use.

3.2.2 Analysis of Selected Members of the Microbiota

The composition of the intestinal microbiome is most commonly assessed by DNA-based approaches, although metabolome, transcriptome, and proteome studies can often be similarly informative [11]. Usually, a marker gene is amplified and the amplicon is then sequenced. Amplicons are commonly between 150 and 500 base pairs in length. The most widely used marker to identify the presence of certain microbial elements in the community of interest is the DNA encoding for the 16S ribosomal RNA subunit (16S rDNA). The following protocol allows for quantifying the amount of specific 16S rDNA in total DNA extracted from fecal material.

1. Choose the PCR primers of interest. These include primer pairs covering total bacteria (such as the widely used pair F8-R338, *see* **Note 7**) as well as those specific for the bacterial taxa of interest. 16S rDNA reference sequences are available in several databases, such as SILVA [12] and the Ribosomal Database Project [13].
2. Mix 5 μL of the sample DNA solution with 5 pmol of each primer.
3. Add 0.2 μL of bacteria-specific probes and 5 μL of universal qPCR mix.
4. Run real-time PCR at 95 °C for 120 s, followed by 40 cycles of 95 °C for 3 s and 64 °C for 30 s.
5. Quantify abundance of bacterial taxa by normalizing to signal obtained for total bacterial 16S sequences (*see* **Note 8**).

3.2.3 Comprehensive Sequencing Analysis of Microbial Communities

The realization that the microbiome is controlled by components of the innate immune system and that the composition of intestinal microbial communities substantially affects many aspects of the host's physiology has precipitated the development of technologies for the fast collection of microbiome profiles at relatively low cost.

Experimental tools for such profiling studies have been reviewed in depth recently [14], and we will focus here on the general framework and basic steps necessary for the determination of the bacterial microbiome of inflammasome-deficient mice by 16S rDNA sequencing.

1. Carefully choose the PCR primers for amplification (*see* **Note 9**).
2. Perform three replicate PCRs for each fecal DNA.
3. Pool the three reactions in an equimolar ratio after DNA quantification with the PicoGreen DNA Kit.
4. Purify the DNA with the Ampure magnetic purification beads.
5. Obtain sequences from isolated bacterial DNA (*see* **Note 10**).
6. After acquiring DNA sequences, the errors that have been produced during the sequencing process need to be removed by clustering the obtained sequences. Useful bioinformatic tools for this step are Denoiser and AmpliconNoise [15].
7. Next, process the reads using bioinformatic tools which allow for comparative analysis of microbial taxa based on targeted amplicon sequences, such as QIIME [16] or mothur [17]. These programs evaluate the amplicons based on sequence similarity and cluster them into operational taxonomic units (OTUs), usually binning sequences sharing at least 97 % nucleotide sequence identity into OTUs.
8. To extract phylogenetic information, reference databases such as SILVA [12] or greengenes [18] can be used and the obtained phylogenies can be interpreted using analysis tools such as UniFrac [19] (*see* **Note 11**).

4 Notes

1. Cohousing should ideally start immediately after weaning, i.e., at about 3 weeks of age, and should be maintained until the experiment is performed. A minimum of 4 weeks of cohousing needs to be strictly kept to avoid partial transmission of the microbiota [20]. Changes of cage bedding and washing of cages can be routinely performed during the cohousing period. Due to their less aggressive behavior, female C57BL/6 mice are generally more suitable for cohousing experiments. However, when cohoused immediately after weaning, male mice can also be used.
2. Separation of previously cohoused mice leads to the successive loss of the "foreign" microflora composition [5], supporting the concept that the host genotype defines the availability of niches for intestinal microbial colonization.

3. In order to increase efficiency of bacterial transfer, mice may be starved for 6 h before gavage. When performing the gavage, hyperextend the neck of the mouse and slowly insert the gavage needle to avoid injury when delivering material into the stomach. Mice should not be anesthetized for this procedure.
4. In our experience, 4 weeks of cohousing is the most reliable way to ensure full equilibration of the microflora between two genotypes of interest. Cross-fostering may be devised to analyze for the effect of maternal transmission to newborns. Oral gavage of fecal homogenates is most efficient when the recipient mouse does not harbor a full complex microbiota, such as germ-free mice or antibiotic-pretreated mice. Otherwise, the indigenous flora provides colonization resistance against incoming bacteria [21]. The latter method is most useful to introduce defined members of the microbial community, while cohousing and cross-fostering will change the entirety of the intestinal microbiota composition over a longer period of time.

   The use of heterozygous breeding is an efficient way of ensuring an equilibrated microbiota composition between inflammasome knockout mice and their wild-type or heterozygous littermates. In this setting, to test for a role of genotype-dependent dysbiosis, inflammasome-deficient mice and their littermates should be separated upon weaning and housed in separate cages for several weeks, allowing for divergent development of microflora compositions based on the genetic makeup of the respective hosts. In the experimental setup of interest, these separately housed animals should be compared to non-separated knockouts and their littermates, to assess whether the spatial separation lead to the establishment of different intestinal microbiota compositions.
5. If bead beating technology is unavailable, extensive vortexing and manual homogenization with a pestle under sterile conditions might provide a suitable alternative.
6. The overnight precipitation may be used as a break point in the protocol. However, in our experience, 1 h at –20 °C is already sufficient for full precipitation efficiency.
7. Primer sequences: F8 5′-AGAGTTTGATCCTGGCTCAG-3′; R338 5′-TGCTGCCTCCCGTAGGAGT-3′.
8. Dysbiosis in inflammasome-deficient mice has been associated with an outgrowth of *Prevotellaceae* and *TM7*, and an under-representation of *Lactobacillaceae* [5]. These taxa might therefore serve as a useful starting point when targeting aberrantly represented members of the bacterial community in inflammasome-deficient mice by qPCR.
9. The available primer pairs vary greatly with respect to compatibility with various sequencing instruments, taxonomic coverage, and the depth of phylogenetic specificity. The above mentioned primer pair F27-R338, for instance, albeit useful

for mapping the entire bacterial community, biases against certain groups [14].

10. To date, several sequencing platforms are available which devise various technological principles, including pyrosequencing (such as the Roche 454 Sequencer, FLX and FLX Titanium), clonal arrays (e.g., HiSeq 2000, Illumina), and capillary sequencing (such as ABI 3730). The individual platforms have different performance features with respect to speed, error rate, cost, and read length.
11. Once assigned to OTUs and analyzed by UniFrac, 16S rDNA sequencing data is still fairly extensive and hard to interpret. To facilitate visual analysis, the data is typically depicted using principal component analysis (PCoA), which identifies common relationships among microbial communities and visualizes them in a two- or three-dimensional space. In a PCoA plot, each community is represented by a dot, with mice with a similar microbiota composition being presented close to each other.

References

1. Martinon F, Burns K, Tschopp J (2002) The inflammasome: a molecular platform triggering activation of inflammatory caspases and processing of proIL-beta. Mol Cell 10(2): 417–426
2. Strowig T, Henao-Mejia J, Elinav E, Flavell R (2012) Inflammasomes in health and disease. Nature 481(7381):278–286
3. Franchi L, Munoz-Planillo R, Nunez G (2012) Sensing and reacting to microbes through the inflammasomes. Nat Immunol 13(4): 325–332
4. Zaki MH, Lamkanfi M, Kanneganti TD (2011) The Nlrp3 inflammasome: contributions to intestinal homeostasis. Trends Immunol 32(4):171–179
5. Elinav E, Strowig T, Kau AL, Henao-Mejia J, Thaiss CA, Booth CJ, Peaper DR, Bertin J, Eisenbarth SC, Gordon JI, Flavell RA (2011) NLRP6 inflammasome regulates colonic microbial ecology and risk for colitis. Cell 145(5):745–757
6. Henao-Mejia J, Elinav E, Jin C, Hao L, Mehal WZ, Strowig T, Thaiss CA, Kau AL, Eisenbarth SC, Jurczak MJ, Camporez JP, Shulman GI, Gordon JI, Hoffman HM, Flavell RA (2012) Inflammasome-mediated dysbiosis regulates progression of NAFLD and obesity. Nature 482(7384):179–185
7. Vijay-Kumar M, Aitken JD, Carvalho FA, Cullender TC, Mwangi S, Srinivasan S, Sitaraman SV, Knight R, Ley RE, Gewirtz AT (2010) Metabolic syndrome and altered gut microbiota in mice lacking toll-like receptor 5. Science 328(5975):228–231
8. Rakoff-Nahoum S, Medzhitov R (2008) Innate immune recognition of the indigenous microbial flora. Mucosal Immunol 1(Suppl 1): S10–S14
9. Mandar R, Mikelsaar M (1996) Transmission of mother's microflora to the newborn at birth. Biol Neonate 69(1):30–35
10. Harris NL, Spoerri I, Schopfer JF, Nembrini C, Merky P, Massacand J, Urban JF Jr, Lamarre A, Burki K, Odermatt B, Zinkernagel RM, Macpherson AJ (2006) Mechanisms of neonatal mucosal antibody protection. J Immunol 177(9):6256–6262
11. Shi Y, Tyson GW, DeLong EF (2009) Metatranscriptomics reveals unique microbial small RNAs in the ocean's water column. Nature 459(7244):266–269
12. Pruesse E, Quast C, Knittel K, Fuchs BM, Ludwig W, Peplies J, Glockner FO (2007) SILVA: a comprehensive online resource for quality checked and aligned ribosomal RNA sequence data compatible with ARB. Nucleic Acids Res 35(21):7188–7196
13. Cole JR, Wang Q, Cardenas E, Fish J, Chai B, Farris RJ, Kulam-Syed-Mohideen AS, McGarrell DM, Marsh T, Garrity GM, Tiedje JM (2009) The Ribosomal Database Project: improved alignments and new tools for rRNA analysis. Nucleic Acids Res 37(Database issue):D141–D145
14. Kuczynski J, Lauber CL, Walters WA, Parfrey LW, Clemente JC, Gevers D, Knight R (2012) Experimental and analytical tools for studying the human microbiome. Nat Rev Genet 13(1):47–58

15. Quince C, Lanzen A, Davenport RJ, Turnbaugh PJ (2011) Removing noise from pyrosequenced amplicons. BMC Bioinformatics 12:38
16. Caporaso JG, Kuczynski J, Stombaugh J, Bittinger K, Bushman FD, Costello EK, Fierer N, Pena AG, Goodrich JK, Gordon JI, Huttley GA, Kelley ST, Knights D, Koenig JE, Ley RE, Lozupone CA, McDonald D, Muegge BD, Pirrung M, Reeder J, Sevinsky JR, Turnbaugh PJ, Walters WA, Widmann J, Yatsunenko T, Zaneveld J, Knight R (2010) QIIME allows analysis of high-throughput community sequencing data. Nat Methods 7(5):335–336
17. Schloss PD, Westcott SL, Ryabin T, Hall JR, Hartmann M, Hollister EB, Lesniewski RA, Oakley BB, Parks DH, Robinson CJ, Sahl JW, Stres B, Thallinger GG, Van Horn DJ, Weber CF (2009) Introducing mothur: open-source, platform-independent, community-supported software for describing and comparing microbial communities. Appl Environ Microbiol 75(23):7537–7541
18. DeSantis TZ, Hugenholtz P, Larsen N, Rojas M, Brodie EL, Keller K, Huber T, Dalevi D, Hu P, Andersen GL (2006) Greengenes, a chimera-checked 16S rRNA gene database and workbench compatible with ARB. Appl Environ Microbiol 72(7):5069–5072
19. Lozupone C, Hamady M, Knight R (2006) UniFrac—an online tool for comparing microbial community diversity in a phylogenetic context. BMC Bioinformatics 7:371
20. Stecher B, Chaffron S, Kappeli R, Hapfelmeier S, Freedrich S, Weber TC, Kirundi J, Suar M, McCoy KD, von Mering C, Macpherson AJ, Hardt WD (2010) Like will to like: abundances of closely related species can predict susceptibility to intestinal colonization by pathogenic and commensal bacteria. PLoS Pathog 6(1):e1000711
21. Stecher B, Hardt WD (2011) Mechanisms controlling pathogen colonization of the gut. Curr Opin Microbiol 14(1):82–91

Chapter 15

Quantification of Adipose Tissue Leukocytosis in Obesity

Ryan Grant, Yun-Hee Youm, Anthony Ravussin, and Vishwa Deep Dixit

Abstract

The infiltration of immune cell subsets in adipose tissue termed "adipose tissue leukocytosis" is a critical event in the development of chronic inflammation and obesity-associated comorbidities. Given that a significant proportion of cells in adipose tissue of obese patients are of hematopoietic lineage, the distinct adipose depots represent an uncharacterized immunological organ that can impact metabolic functions. Here, we describe approaches to characterize and isolate leukocytes from the complex adipose tissue microenvironment, to aid mechanistic studies to better understand the role of specific pattern recognition receptors (PRRs) such as inflammasomes in adipose-immune cross talk.

Key words B cell, T cell, Macrophage, Granulocyte, Flow cytometry, Adipose tissue digestion

1 Introduction

Adipose tissue is a complex and diverse cellular organ that is primarily involved in the regulation of energy homeostasis. Adipocytes are the predominant cell type in adipose depots and control metabolism by storing excess calories as lipids together with production of endocrine hormones such as leptin and adiponectin [1]. The capacity of adipose tissue to expand and store energy as lipids represents a critical adaptation to chronic caloric excess. In addition to adipocytes, adipose tissue is composed of stromal-vascular fraction (SVF) cells and cells of the hematopoietic lineage (*see* Fig. 1). It is now established that adipose tissue expansion in obesity is associated with an increase in macrophages, neutrophils, T cells, B cells, and mast cells in adipose tissue [2–7] (*see* Fig. 1). Thus, "adipose leukocytosis," or increased infiltration of leukocytes into adipose tissue during obesity, represents an important link between adaptive adipose tissue remodeling in response to energy excess and the emergence of chronic inflammation-associated insulin resistance.

The enzymatic dispersion of adipose tissue and subsequent processing yields adipocytes (floating fraction) and a cell pellet called SVF [2, 7, 8] (*see* Fig. 1). In obesity, depending on specific

Christine M. De Nardo and Eicke Latz (eds.), *The Inflammasome: Methods and Protocols*, Methods in Molecular Biology, vol. 1040, DOI 10.1007/978-1-62703-523-1_15, © Springer Science+Business Media New York 2013

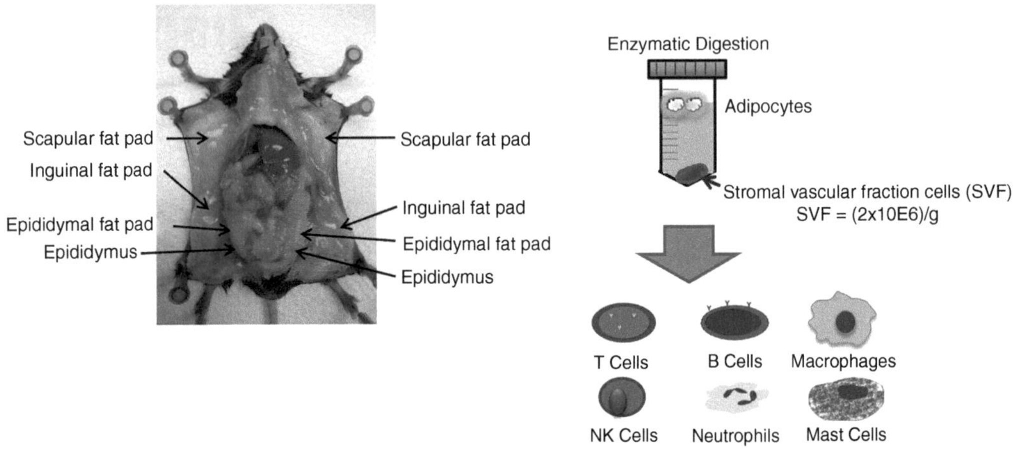

Fig. 1 Overview of the separation and analysis of leukocytes from adipose tissue. Mouse adipose tissue is collected and weighed prior to subsequent analysis. Adipose tissue is digested to yield a stromal-vascular fraction (SVF). The SVF contains immune cell subsets including T and B cells, macrophages, NK cells, neutrophils, and mast cells, which are increased by obesity

sites, the adipose depots can contain ~2 to 5 million SVF cells/g and roughly 50–75 % of these cells can be of hematopoietic lineage [7, 9] (*see* Fig. 2). Importantly, in morbid or extreme obesity in humans, the total adipose tissue mass can increase to constitute up to 50 % of the total body mass. Thus, the expanded adipose tissue represents a largely uncharacterized immunological organ with distinct leukocytes with potentially unique function to regulate immune-metabolic cross talk.

The control of caspase-1 activation and subsequent production of active IL-1β, IL-18, and other yet to be identified proteins is an integral part of innate immunity. Caspase-1 is activated by multiprotein scaffolding complexes termed "inflammasomes," which are responsive to both exogenous pathogen associated molecular patterns (PAMPs), and damage associated molecular patterns (DAMPs) [10]. Even though IL-1β and IL-18 play key roles in innate immunity, they also cause host tissue damage, especially during chronic inflammation. These cytokines have pleiotropic roles to alter the function of lymphocytes and other immune cells, as well as nonimmune cells including adipocytes and hepatocytes; thus, it is important to be able to distinguish their effects on different cell types in vivo. This may be especially important when analyzing the influence of the inflammasomes on metabolic organs in chronic diseases like obesity and type 2 diabetes [9, 11]. Through the use of sequential enzymatic digestions to disperse adipose depots and multicolor flow cytometry, we have shown that specific leukocyte subsets including T cells and macrophages are affected by NLRP3 ablation in the adipose tissue during high-fat feeding [9]. Reduction in adipose tissue inflammation is correlated with lower M1 macrophage numbers and effector T cell population together with increased whole body insulin sensitivity [9].

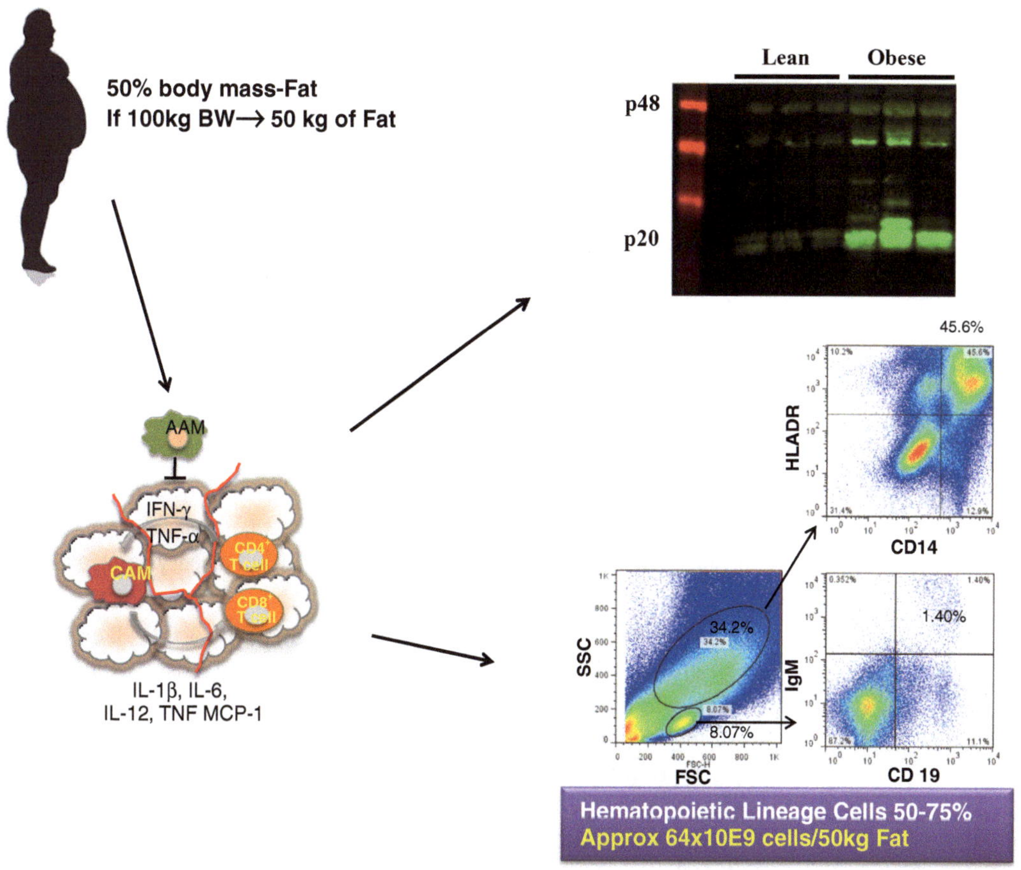

Fig. 2 Overview of the analysis of leukocytes from adipose tissue. Human adipose tissue is collected, weighed and digested prior to downstream analysis. FACS plots from human adipose tissue SVF taken from a female subject (BMI 29.9) showing $CD14^+HLADR^+$ macrophages and IGM^+CD19^+ B cells. Adipose tissue leukocytosis and the accretion of body fat lead to a large increase in the total amount of cells in the fat ~64×10^9 cells/50 kg fat. Western blot analysis of caspase-1 activation (p20) in visceral adipose tissue from 9-month WT lean, WT diet-induced obese (DIO) mice

Other studies have borne out the specific effects of IL-1β and IL-18 on adipocytes and hepatocytes [12, 13].

Analysis of adipose tissue immune cell subsets is difficult because of the complex structure of adipose tissue. Care must be taken with interpretation of data generated from both isolated SVF and floating adipocyte fractions. In addition to leukocytes, the SVF also contains mesenchymal and endothelial cells. For leukocyte analysis, immune cell subsets must be liberated from the tissue, without disruption of their cell surface receptors and then subsequently analyzed. The floating adipocyte fraction is not purely adipocytes because macrophages not only become larger, but more buoyant as they engulf lipids. Thus, the floating adipocyte fraction is contaminated with large buoyant macrophages as demonstrated in Fig. 3 showing CD11c expression in both the stromal-vascular

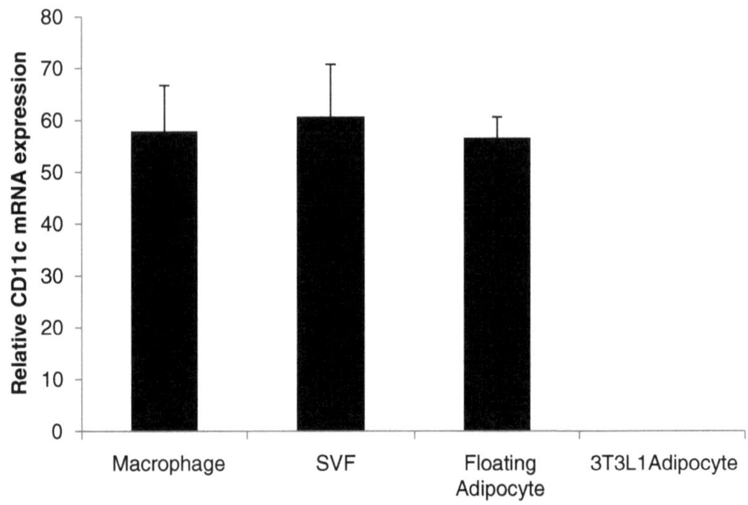

Fig. 3 Relative mRNA expression of CD11c in F4/80$^+$ adipose tissue macrophages, SVF and floating adipocyte fractions and 3T3-L1 adipocytes (mean ± SEM)

and floating adipocyte fractions. Detection of myeloid cell expressed cytokines and pattern recognition receptors such as NLRP3 in the adipocyte fraction can thus be confounded by the presence of contaminating lipid engorged adipose tissue macrophages.

Herein, we describe methods for collection and digestion of adipose tissue, in order to isolate SVF and adipocyte cell populations that may be relevant to future studies on the mechanisms underlying immune-metabolic interactions in obesity-associated chronic diseases. Subsequent to immune cell isolation, we describe analysis of specific immune cell populations in adipose tissue including the surface staining for specific marker proteins and gating strategies useful for flow cytometric analysis.

2 Materials

2.1 Collection of Subcutaneous and Visceral Adipose Tissue Depots

1. Sterile forceps and scissors.
2. Culture medium: RPMI 1640 supplemented with 5 % fetal bovine serum (FBS), 1 % penicillin/streptomycin.
3. Sterile 24-well plate.
4. 70 % ethanol.

2.2 Digestion of Fat

1. Sterile scissors.
2. 70 % ethanol.
3. Paper towels.
4. Collagenase type I buffer: Add 0.2 g of collagenase type 1 powder to 200 mL of Hank's Balanced Salt Solution (HBSS) without Ca^{2+} and Mg^{2+}, pH 7.1. Filter through a 22 μm filter.

5. 22 μm filter.
6. Ice.
7. 6-well plate.
8. Sterile transfer pipettes.
9. 3 × 50 mL conical tube.
10. Shaking water bath at 37 °C.
11. Laminar flow hood.
12. Centrifuge at 4 °C.
13. Ammonium-Chloride-Potassium (ACK) Lysing Buffer.
14. Sterile culture medium (*see* Subheading 2.1).
15. Sterile 100 μm cell strainer.
16. Sterile 40 μm cell strainer.
17. Phosphate buffered saline (PBS).
18. Bovine serum albumin (BSA).
19. Hemocytometer.

2.3 FACS Staining

1. 5 mL round bottom tubes.
2. Stain buffer: dPBS, pH 7.4, 2 % FBS and 0.09 % NaN_3.
3. FC-Block (purified anti-CD16/CD32).
4. Vortex.
5. Aluminum foil.
6. Ice.
7. PBS.
8. Fixative: 1 % paraformaldehyde.
9. Fluorescently labeled antibodies (*see* Table 1 (mouse)).

3 Methods

3.1 Collection of Subcutaneous and Visceral Mouse Adipose Tissue Depots

1. Place enough cold sterile culture medium into the wells of a sterile 24-well plate so that when the tissue is collected, it is covered in media (~1 mL) (*see* **Note 1**).
2. After the mouse is euthanized, spray the body with 70 % ethanol.
3. Make an incision through the skin and peritoneum from the lower stomach to rib cage. Continue the incision out along the four limbs and separate the peritoneum from the skin (*see* **Note 2**).
4. The epididymal fat pad is located adjacent to the epididymis (*see* Fig. 1), and should be removed carefully so that adipose tissue is not contaminated by other tissue. Place tissues into the media in the prepared 24-well plate (*see* **Note 3**).

Table 1
Antibodies used for analysis of immune cell subsets in adipose tissue

FSC/SSC gate	Cell population	Cell subpopulation	Antibody	Mouse clone
Lymphocyte				
	T cell		CD3	145-2C11
	CD4⁺ T cell		CD4	RM4-5
		Naïve	$CD44^{-}CD62L^{+}$	IM7; Mel-14
		Effector memory	$CD44^{+}CD62L^{-}$	IM7; Mel-14
	$CD8^{+}$ T cell		CD8	53-6.7
		Naïve	$CD44^{-}CD62L^{+}$	IM7; Mel-14
		Effector memory	$CD44^{+}CD62L^{-}$	IM7; Mel-14
	B cell		B220	RA-6B2
		Mature B cells	IGM,CD19	11/41; ebio1D3
	Granulocyte		Grl	RB6-8C5
Macrophage				
	Macrophage		F4/80	BM8
		M1	CD11c	N418
		M2	CD206	C068C2
	Granulocyte		Grl	RB6-8C5

5. The inguinal (subcutaneous) fat pads lie between the peritoneum and the skin (*see* Fig. 1). Remove the fat pad by lifting it up and cutting along the skin. This fat pad is fairly large and continues around to the back close to the spinal column (*see* **Note 4**).
6. Weigh fat pads (*see* **Note 5**).

3.2 Digestion of Fat

1. Prepare collagenase type I buffer as in **item 4** of Subheading 2.2, and place on ice.
2. Perform the following steps in a laminar flow hood and on ice. Transfer the fat tissue to a 6-well plate and cover the tissue with collagenase type 1 buffer using a sterile transfer pipette (*see* **Notes 6** and 7).
3. Use two scissors to cut the tissue until it is in small pieces (*see* **Note 8**).
4. Transfer the cut tissue and collagenase buffer to a labeled 50 mL conical tube using a sterile transfer pipette (*see* **Note 9**).

5. Add more collagenase buffer to the tubes until it reaches at least 10 mL and is at a 1:1 ratio of adipose tissue to collagenase buffer (*see* **Note 10**).
6. Place the 50 mL tube on ice until all samples have been prepared.
7. Spray scissors and forceps with 70 % ethanol and wipe with a paper tissue before mincing the next tissue sample. Change sterile transfer pipettes between samples.
8. Once all samples are minced and at the appropriate volume, place them into a rack and incubate the tissue in a shaking water bath at 37 °C for 1 h (*see* **Notes 11** and **12**).
9. Vortex the tubes every 10 min during incubation, to ensure that the collagenase has full access to the tissue (*see* **Note 13**).
10. Centrifuge the tubes at 500×*g* for 5 min at 4 °C.
11. In the laminar flow hood, remove the supernatant, and resuspend and disperse the pellet in 2 mL of ACK Lysing Buffer for 2 min at room temperature to remove red blood cells from the pellet.
12. After 2 min, neutralize the reaction by adding 5 mL of culture medium.
13. Filter the cells and media using a sterile 100 μm cell strainer into a new, labeled 50 mL conical tube. Discard the filter and old tubes when done.
14. Centrifuge the tubes at 500×*g* for 5 min at 4 °C.
15. Discard the supernatant.
16. Wash the cell pellet by adding 5–10 mL of PBS. Disperse the cell pellet using a sterile transfer pipette by pipetting up and down several times.
17. Filter the sample again, this time using a 40 μm cell strainer into a new 50 mL conical tube.
18. Centrifuge the tubes at 500×*g* for 5 min at 4 °C.
19. Discard the supernatant (*see* **Note 14**).
20. Add 1 mL of PBS with 0.1 % BSA or culture medium. Disperse the cells using a sterile transfer pipette and place on ice.
21. Count the stromal-vascular fraction cells using a hemocytometer and continue downstream processing (*see* **Note 15**).

3.3 FACS Staining

1. Transfer the desired number ($1–2\times10^6$) of stromal-vascular cells and splenocyte controls to a labeled 5 mL round bottom FACS tube and centrifuge at 500×*g* for 5 min at 4 °C (*see* **Notes 16** and **17**).
2. Discard the supernatant.
3. Add 1 mL of stain buffer and disperse the cells.
4. Centrifuge at 500×*g* for 5 min at 4 °C.

5. Discard the supernatant.
6. Add 50 μL of stain buffer and 2 μL FC-Block to each tube. Disperse the pellet.
7. Make a master mix that includes all of the antibodies that will be used for the FACS analysis. Table 1 lists antibodies that can be used to analyze specific immune cell populations. Add 2 μL of antibody per sample to the master mix, typical antibody concentrations from the manufacturer are 0.2 mg/mL, but the concentration can vary from supplier to supplier (*see* **Note 18**).
8. Add 2 μL of each antibody per tube. If using four different antibodies, then 8 μL of master mix will be added (*see* **Note 19**).
9. Vortex all the tubes.
10. Cover tubes with foil and incubate for 45 min on ice (*see* **Note 20**).
11. Add 500 μL of PBS to each tube.
12. Centrifuge tubes at 500 × *g* for 5 min at 4 °C.
13. Discard supernatant.
14. Resuspend and disrupt the cell pellet in 750 μL of PBS.
15. Centrifuge tubes at 500 × *g* for 5 min at 4 °C.
16. Discard supernatant.
17. Wash an additional time by repeating **steps 14–16** of Subheading 3.3.
18. After discarding the supernatant, resuspend the cells in 500 μL of 1 % paraformaldehyde.
19. Cover the tubes with aluminum foil to protect from light and leave them on ice or in a 4 °C refrigerator until they will be analyzed.
20. Run the FACS analysis (*see* **Notes 21** and **22**).

3.4 FACS Gating Strategy and Analysis

All the FACS data should be analyzed by post collection compensation using appropriate analysis software (FlowJO, Treestar Inc). Resources for FlowJo analysis can be found at http://www.flowjo.com/home/tutorials/.

1. Gating strategy: Compared to the spleen, adipose tissue is much more complex, displaying a more varied forward and side scatter with a higher incidence of large dense cells that appear towards the upper right quadrant of the forward and side scatter FACS plot (*see* Fig. 4). Lymphocytes reside in a small population of cells that appear in the lower left quadrant of the forward and side scatter, and the spleen has a much higher frequency of these cells compared with adipose tissue, as would be expected (*see* Fig. 4). Figure 5 displays the gating for adipose tissue lymphocytes and macrophages. B cells and T

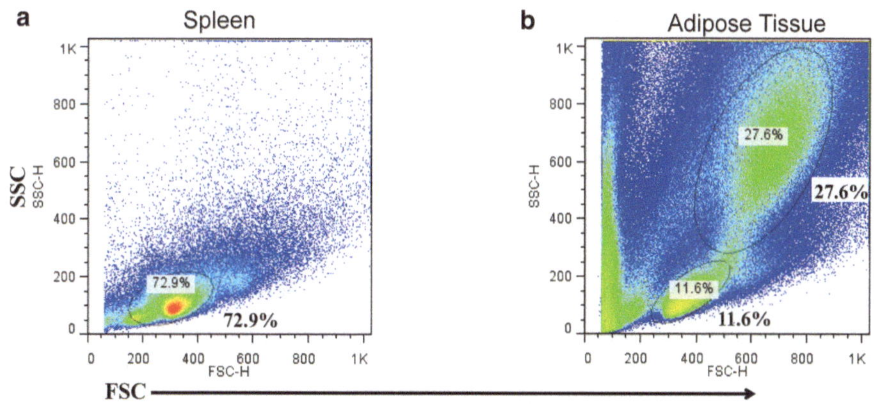

Fig. 4 Comparison of forward scatter and side scatter FACS plots from (**a**) splenocytes and (**b**) SVF cells derived from epididymal fat pad of a 4-month-old obese wild type mouse

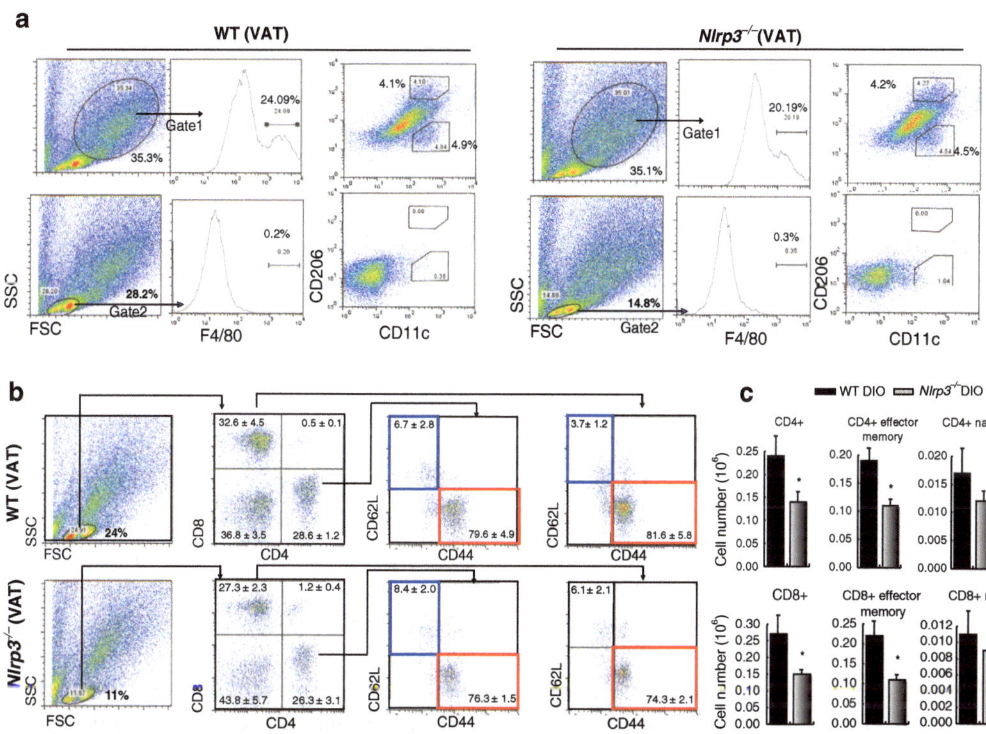

Fig. 5 Gate placement for analysis of macrophages and lymphocytes, and the influence of NLRP3 ablation on T cell subpopulations in visceral adipose tissue (VAT) during died-induced obesity (DIO). (**a**) *Upper row* of FACS plots demonstrates that macrophages reside in a population of large SVF cells (Gate 1), while the *lower row* demonstrates that they do not reside in the small "lymphoid" gate (Gate 2) in VAT of 9-month WT and *Nlrp3*$^{-/-}$ DIO mice. Starting with the FSC and SSC (*left*) populations are sequentially gated for macrophage analysis, first by F4/80 expression and then by CD11c (M1 marker) and CD206 (M2 marker). (**b**) Sequential analysis of lymphocytes starting with Gate 2 (lymphoid gate), then separating lymphocytes by CD4$^+$ and CD8$^+$ T cells, and then evaluating the naïve (CD62L$^+$CD44$^-$, *blue boxes*) and effector memory (CD62L$^-$CD44$^+$, *red boxes*) T cells in both the CD4$^+$ and CD8$^+$ populations. (**c**) Absolute numbers (in million cells) of naïve and effector CD4$^+$ and CD8$^+$ T cells in adipose tissue of 9 month old WT-DIO and *Nlrp3*$^{-/-}$ DIO mice (Figure adapted from Vandanmagsar et al., Nat. Med. 10, 179–188: 2011)

cells are located within the "lymphocyte gate," while macrophages are larger and display more variation in size and density, so the gated population is larger for these cells (*see* Fig. 5). Granulocytes consist of neutrophils, eosinophils, and basophils, and the size range of these cells varies from just larger than lymphocytes, to cells the size of macrophages. Thus, granulocytes are present in both the "lymphocyte gate" and the "macrophage gate." For mouse adipose tissue, a representative set of gates can be made for the targeted immune cell populations and applied to all adipose tissue samples.

2. Once the gates have been set and applied to all samples, the samples are ready for statistical analysis. Analyze the cell frequency starting with the FSC x SSC gated population and continue on through all the gates created. The number of leukocytes in distinct adipose depots can be calculated as shown in the example below:
   - Example using population percentages typical for mice: Lymphocyte gate 12 %, B cell ($B220^+$) frequency 15 %, mature B cells (IGM^+,$CD19^+$) 90 %, 1.5 g adipose tissue mature B cells/g adipose tissue = (Total cell counts × 12 % × 15 % × 90 %)/1.5 g adipose tissue.

3.5 Digestion and Analysis of Human Adipose Tissue

Human subcutaneous and visceral adipose tissue samples obtained from biopsies or from liposuction materials can be used to analyze adipose tissue immune cell subsets (*see* **Note 23**).

1. Weigh biopsy samples (*see* **Note 5**).
2. Proceed with tissue digestion as described in Subheading 3.2.
3. Proceed with antibody staining protocols for flow cytometric analysis as described in Subheading 3.3.
4. Analyze human samples in the FlowJo software (or other flow cytometry analysis software) by following the same steps as described for the mouse samples. Proceed as outlined in Subheading 3.4. Even though the FSC x SSC plots look different, the general location of the immune cell populations in the plots are similar (mouse: *see* Fig. 5, human: *see* Figs. 2 and 6). Between-subject variation is much higher for the human adipose tissue immune cell subsets (*see* Fig. 6). Each subject should be analyzed individually and therefore a representative set of gates should not be used.
5. It is necessary to combine FACS analysis with measures of caspase-1 activation to fully understand the impact of inflammasome activation on adipose tissue biology (*see* **Note 24**).

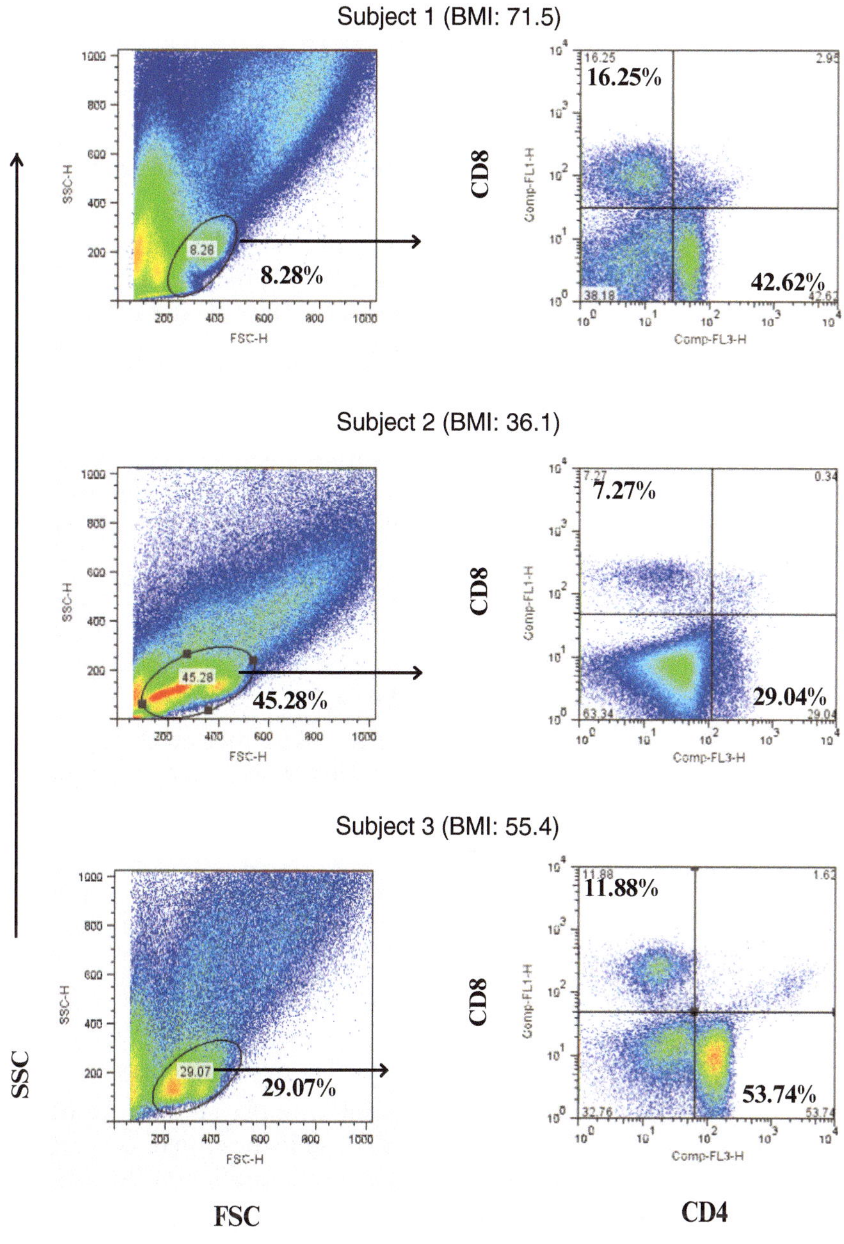

Fig. 6 Human subjects display variability in adipose tissue T cell populations. FACS analysis of SVF-T cells from omental adipose tissue obtained from obese patients undergoing bariatric surgery. Lymphocyte populations (*left*) were gated and then analyzed for CD4+ and CD8+ T cells (*right*). Subjects were all female with a BMI (from *top* to *bottom*) of 71.5, 36.1, 55.4, respectively

4 Notes

1. Medium and tissue should be cold and kept on ice throughout the protocol.
2. The skin and limbs may be pinned down as in Fig. 1 to aid in the dissection.
3. Care should be taken to minimize the amount of time between removal of fat pads and placing them in media. This is important to prevent drying out of the cells and to keep the temperature in the tissue cold.
4. Carefully look over the inguinal fat pad for lymph nodes. These are embedded within the fat pad, but may not be easily visible, offering only slight discoloration. Once spotted, cut adjacent to the lymph node. Then, using forceps and downward pressure, grasp the tissue under the lymph node and cut underneath to remove it.
5. Fat pads should be blotted on a paper towel prior to weighing to remove any excess moisture, and ensure an accurate weight measurement. Place fat pads back into media on ice and then proceed to Subheading 3.2.
6. Concentration of collagenase buffer needs to be at 0.1 % and at least 10 mL will be needed per depot, per mouse.
7. A suspension culture 6-well plate is useful to limit adherence of cells to the plate bottom.
8. To determine how finely the tissue is minced, use forceps to check for larger pieces of tissue, which can then be cut into smaller pieces.
9. The transfer pipette tip can be cut off to allow larger pieces of tissue to pass through. If tissue does not easily pass through the pipette opening, further mincing is necessary.
10. The adipose to collagenase buffer ratio can be determined by eye and is easily visible. A ratio lower than 1:1 (collagenase–tissue) will not allow complete digestion of the tissue.
11. For the incubation, a shaking water bath is the optimal method because the water provides very efficient heat transfer to the samples. However, if a shaking water bath is not available, a cell culture incubator at 37 °C can be used as an alternative. If a cell incubator will be used, wrap the tubes and rack in aluminum foil to ensure even heat distribution.
12. Do not digest tissue for more than 1 h, because this can lead to increased cell death and lower cellular yield. If digestion is incomplete, in future experiments a higher collagenase buffer to tissue ratio can be used, or a higher concentration of collagenase in HBSS can be prepared.

13. Vortexing is necessary because the fat tends to form a floating mass at the top of the collagenase buffer during the digestion.
14. Floating adipocytes can be collected from the supernatant above the pellet if needed.
15. Isolation of specific cell populations using commercially available kits for untouched immune cell populations designed for lymphoid organs (bone marrow and spleen) do not provide pure populations. Although the kits remove other immune cells, stromal-vascular cells remain. Positive selection using a bead-based system or cell sorting is required.
16. For each FACS experiment, control samples used for gating and compensation should be prepared. A tube of unstained splenocytes and splenocytes stained with each antibody alone are used for unstained and single color controls and will be necessary for compensation of the individual channels. It may be useful to run each splenocyte sample as a comparator to the adipose tissue stromal-vascular fraction (SVF) samples. The larger SVF cells display high autofluorescence (presumably due to presence of lipid and/or particulate matter) and staining requires appropriate isotype IgG controls.
17. Stain between 1 and 2×10^6 cells for collection of 500,000 to 1×10^6 events. Using FC-Block and performing wash steps reduces nonspecific antibody binding to cells.
18. The number of antibodies used will depend on the number of detectors on the flow cytometer used and the cell populations analyzed. Be sure to add only 2 μL of the appropriate antibody to single color controls.
19. Be sure the pipette tip touches the fluid to ensure that the complete volume of antibody reaches the fluid. Change pipette tips each time.
20. Keeping the tubes cold keeps the receptor sites at the cell surface and results in better staining.
21. For ease of analysis, order the samples with the unstained control first, followed by the single color controls, which are then followed by the samples that will be analyzed.
22. Useful references for the frequency of adipose tissue immune cell subset include Winer et al. Nature Medicine, Nishimura et al. Nature Medicine, Yang et al. Journal of Immunology, Vandanmagsar et al. Nature Medicine, Lumeng et al. Journal of Clinical Investigation [3, 9, 14–16].
23. A caveat of using liposuction material is significant contamination of blood derived leukocytes.
24. Analysis of inflammasome activation: Although FACS analysis is necessary to characterize populations of immune cells in

adipose tissue, it must be combined with other analysis to understand how inflammasome activation is affecting the local adipose tissue environment. Western blots of frozen adipose tissue, or isolated macrophages can be used to analyze caspase-1 activation and its products. Caspase-1 is a zymogen, with a pro-form at 45 kDa, and after cleavage an active p20 subunit (*see* Fig. 2). IL-1β is also stored in an inactive 31 kDa pro-form and cleaved by caspase-1 to an active p17 subunit.

Acknowledgments

Research in the Dixit lab is supported in part by NIH grants AG31797, DK090556 and the Pennington Foundation. RWG is supported by NIHT32DK064584-10S1. The present work utilized the facilities of the Genomics and CBB Core facilities supported by Pennington Center of Biomedical Research Excellence (NIH 8P20 GM103528) and Nutrition and Obesity Research Center (NIH P30 DK072476).

References

1. Greenberg AS, Obin MS (2006) Obesity and the role of adipose tissue in inflammation and metabolism. Am J Clin Nutr 83(2):461S–465S
2. Weisberg SP, McCann D, Desai M, Rosenbaum M, Leibel RL, Ferrante AW Jr (2003) Obesity is associated with macrophage accumulation in adipose tissue. J Clin Invest 112(12):1796–1808. doi:10.1172/JCI19246
3. Winer DA, Winer S, Shen L, Wadia PP, Yantha J, Paltser G, Tsui H, Wu P, Davidson MG, Alonso MN, Leong HX, Glassford A, Caimol M, Kenkel JA, Tedder TF, McLaughlin T, Miklos DB, Dosch HM, Engleman EG (2011) B cells promote insulin resistance through modulation of T cells and production of pathogenic IgG antibodies. Nat Med 17(5):610–617. doi:10.1038/nm.2353
4. Rausch ME, Weisberg S, Vardhana P, Tortoriello DV (2008) Obesity in C57BL/6J mice is characterized by adipose tissue hypoxia and cytotoxic T-cell infiltration. Int J Obes (Lond) 32(3):451–463. doi:10.1038/sj.ijo.0803744
5. Elgazar-Carmon V, Rudich A, Hadad N, Levy R (2008) Neutrophils transiently infiltrate intra-abdominal fat early in the course of high-fat feeding. J Lipid Res 49(9):1894–1903. doi:10.1194/jlr.M800132-JLR200
6. Liu J, Divoux A, Sun J, Zhang J, Clement K, Glickman JN, Sukhova GK, Wolters PJ, Du J, Gorgun CZ, Doria A, Libby P, Blumberg RS, Kahn BB, Hotamisligil GS, Shi GP (2009) Genetic deficiency and pharmacological stabilization of mast cells reduce diet-induced obesity and diabetes in mice. Nat Med 15(8): 940–945. doi:10.1038/nm.1994
7. Strissel KJ, DeFuria J, Shaul ME, Bennett G, Greenberg AS, Obin MS (2010) T-cell recruitment and Th1 polarization in adipose tissue during diet-induced obesity in C57BL/6 mice. Obesity (Silver Spring) 18(10):1918–1925. doi:10.1038/oby.2010.1
8. Lumeng CN, Deyoung SM, Saltiel AR (2007) Macrophages block insulin action in adipocytes by altering expression of signaling and glucose transport proteins. Am J Physiol Endocrinol Metab 292(1):E166–E174. doi:10.1152/ajpendo.00284.2006
9. Vandanmagsar B, Youm YH, Ravussin A, Galgani JE, Stadler K, Mynatt RL, Ravussin E, Stephens JM, Dixit VD (2011) The NLRP3 inflammasome instigates obesity-induced inflammation and insulin resistance. Nat Med 17(2):179–188. doi:10.1038/nm.2279
10. Petrilli V, Dostert C, Muruve DA, Tschopp J (2007) The inflammasome: a danger sensing complex triggering innate immunity. Curr Opin Immunol 19(6):615–622. doi:10.1016/j.coi.2007.09.002
11. Youm Y-H, Kanneganti T-D, Vandanmagsar B, Zhu X, Ravussin A, Adijiang A, Owen John S, Thomas Michael J, Francis J, Parks John S, Dixit Vishwa D (2012) The NLRP3 inflammasome promotes age-related thymic demise and immunosenescence. Cell Rep 1(1):56–68

12. Stienstra R, Joosten LA, Koenen T, van Tits B, van Diepen JA, van den Berg SA, Rensen PC, Voshol PJ, Fantuzzi G, Hijmans A, Kersten S, Muller M, van den Berg WB, van Rooijen N, Wabitsch M, Kullberg BJ, van der Meer JW, Kanneganti T, Tack CJ, Netea MG (2010) The inflammasome-mediated caspase-1 activation controls adipocyte differentiation and insulin sensitivity. Cell Metab 12(6):593–605. doi:10.1016/j.cmet.2010.11.011
13. Nov O, Kohl A, Lewis EC, Bashan N, Dvir I, Ben-Shlomo S, Fishman S, Wueest S, Konrad D, Rudich A (2010) Interleukin-1beta may mediate insulin resistance in liver-derived cells in response to adipocyte inflammation. Endocrinology 151(9):4247–4256. doi:10.1210/en.2010-0340
14. Yang H, Youm YH, Vandanmagsar B, Ravussin A, Gimble JM, Greenway F, Stephens JM, Mynatt RL, Dixit VD (2010) Obesity increases the production of proinflammatory mediators from adipose tissue T cells and compromises TCR repertoire diversity: implications for systemic inflammation and insulin resistance. J Immunol 185(3):1836–1845. doi:10.4049/jimmunol.1000021
15. Nishimura S, Manabe I, Nagasaki M, Eto K, Yamashita H, Ohsugi M, Otsu M, Hara K, Ueki K, Sugiura S, Yoshimura K, Kadowaki T, Nagai R (2009) $CD8^+$ effector T cells contribute to macrophage recruitment and adipose tissue inflammation in obesity. Nat Med 15(8):914–920. doi:10.1038/nm.1964
16. Lumeng CN, Bodzin JL, Saltiel AR (2007) Obesity induces a phenotypic switch in adipose tissue macrophage polarization. J Clin Invest 117(1):175–184. doi:10.1172/JCI29881

Chapter 16

In Vivo Evaluation of Neutrophil Recruitment in Response to Sterile Particulates

Tamiko Yanagida, Kaoru Orihashi, and Hajime Kono

Abstract

Sterile particulates such as monosodium urate crystals induce inflammasome activation resulting in activation of caspase-1, secretion of IL-1α, and processing of IL-1β. Local production and activation of IL-1 leads to neutrophil recruitment in vivo. Here we describe two quick and simple methods for the evaluation of neutrophil recruitment in the peritoneal cavity and skin in response to sterile particulates, which are dependent on IL-1 receptor signaling.

Key words Neutrophils, Monocytes, IL-1, Peritoneum, Skin, Acute inflammation, Myeloperoxidase

1 Introduction

Recent developments in molecular mechanisms of posttranscriptional processes of IL-1β have led to interest in its physiological activities in vivo. IL-1α and β are cytokines with pleiotropic functions in innate and adaptive immunity, tissue homeostasis, and degenerative as well as regenerative processes [1]. Furthermore, IL-1 cytokines mediate the pathological conditions seen in autoinflammatory diseases and IL-1 inhibition results in a significant reduction in disease severity [2, 3]. The familial cold autoinflammatory syndrome (FCAS) is the prototype of autoinflammatory diseases, which is characterized by fever, arthralgia, and erythematoid macular-like skin eruptions. Mutation of the NLRP3 protein leads to spontaneous inflammasome activation after cold exposure of a percentage of FCAS [4]. The developing exanthema is characterized by an infiltration of neutrophils [5]. This suggests that the acute and direct physiological response to activation of the inflammasome is followed by neutrophil infiltration in situ.

One of the physiologically important phenomena of IL-1-mediated inflammation is the recruitment of neutrophils in response to dead cells and sterile particulates including urate, silica, and cholesterol crystals [6, 7]. These immune responses are

Christine M. De Nardo and Eicke Latz (eds.), *The Inflammasome: Methods and Protocols*, Methods in Molecular Biology, vol. 1040, DOI 10.1007/978-1-62703-523-1_16, © Springer Science+Business Media New York 2013

medically relevant because they contribute to common diseases, such as crystal-induced arthritis. Furthermore, it is believed that IL-1-mediated inflammation contributes to the severity of ischemic diseases. Gout is a disease caused by the formation of monosodium urate crystals in affected joints [8]. This acute arthritis is also characterized by neutrophilic inflammation and ameliorated by recombinant IL-1 receptor antagonists [9]. In this chapter we describe two quick and simple methods for evaluating neutrophil recruitment to the peritoneal cavity or into the skin. The number of neutrophils and monocytes recruited to the peritoneal cavity in response to sterile particulates can be directly evaluated by flow cytometry. In the skin assay described below, neutrophil recruitment is examined by myeloperoxidase (MPO) activity of the tissue following injection of sterile irritants. MPO is an enzyme that is abundantly found in azurophilic granules of neutrophils, and its activity correlates with the number of neutrophils present in the tissue sample [10]. These methods can be applied not only for studying sterile particulate-induced inflammation, but can also serve to investigate inflammation in response to necrotic cell death or towards other innate immune stimuli, such as chemokines or Toll like receptor ligands [11].

2 Materials

2.1 Skin Assay

1. Mice.
2. Razor blade or pet hair clipper.
3. Isoflurane (VetOne).
4. Insulin syringe equipped with a $29G \times 1/2$ needle.
5. Marker (felt-tip pen).
6. 8 mm biopsy punch (Kai industries co., ltd., BP-80F).
7. 12×75 mm, 5 mL polystyrene round bottom test tube (BD # 352003).
8. Polytron tissue homogenizer.
9. Ultrasound sonicator.
10. Colorimetric plate reader (OD 450 and 560 nm).
11. 96-well flat bottom cell culture plates.
12. BCA Protein Assay Kit (Thermo Scientific).
13. MPO Buffer: 50 mM Na_2HPO_4, pH 5.4, 0.5 % Hexadecyl Trimethyl Ammonium Bromide (HTAB, Sigma), 10 mM EDTA (*see* **Note 1**).
14. Assay buffer: 1.67 mg/mL *O*-dianisidine dihydrochloride (Sigma D3252) in 50 mM Na_2HPO_4, pH 5.4 (*see* **Note 2**).
15. Development solution: 3 μL of 31.1 % H_2O_2 added to 10 mL 50 mM Na_2HPO_4, pH 5.4.

2.2 Peritoneal Assay

1. Mice.
2. Small surgical scissors.
3. Insulin syringe with 29G × 1/2 needle.
4. 10 mL syringe with 27G needle and 18G needle.
5. 1.5 mL tubes.
6. Centrifuge.
7. Lavage solution: RPMI1640 with 2 % FCS containing 3 mM EDTA and 10 U/mL heparin.
8. Staining solution: PBS with 2 % FCS and 0.1 % sodium azide.
9. mAbs: Ly-6G-FITC (clone 1A8, BD Biosciences), Ly-6B.2-biotin (clone 7/4, AbD Serotec), APC-Streptavidin (BD Biosciences), anti FcγRIIB/III receptor (clone 2.4G2, BD Biosciencess) (*see* **Note 3**).
10. 7-Amino-actinomycin D (7-AAD) (Calbiochem).
11. Cell counting chamber.
12. Flow cytometer with automatic sampler (e.g., High Throughput Sampler, BD Biosciences).
13. CellQuest software for data acquisition (BD Biosciences).
14. FlowJo software for data analysis (Tree Star Inc.).

3 Methods

Ensure that all animal procedures are carried out by adequately trained persons and that all experiments have received prior ethics approval according to the institutions local regulations.

3.1 Skin Assay

3.1.1 Inducing Inflammation and Collecting the Inflamed Skin Tissue

1. Using a razor blade or animal shaver, shave the mouse peritoneum and dorsal side of the trunk (*see* Fig. 1), at least 1 day before the injection (*see* **Note 4**).
2. Anesthetize the mouse using Isoflurane (*see* **Note 5**).
3. Inject six sites intradermally with 50 μL of sample (irritant of interest), creating small blebs on the skin. Outline the injection areas with a marker (*see* **Note 6** and Fig. 2).
4. 4–24 h after injection, euthanize the mice by a method that is in accordance with the institutions local animal regulations (*see* **Note 7**).
5. Excise the marked skin "blebs" using a biopsy punch and scissors (*see* **Note 8**). Collect each detached bleb containing the inflamed subcutaneous tissue into separate 5 mL flow cytometry tubes.
6. Weigh the skin samples. Proceed to extracting myeloperoxidase (MPO) as in Subheading 3.1.2 or store samples at −80 °C until use.

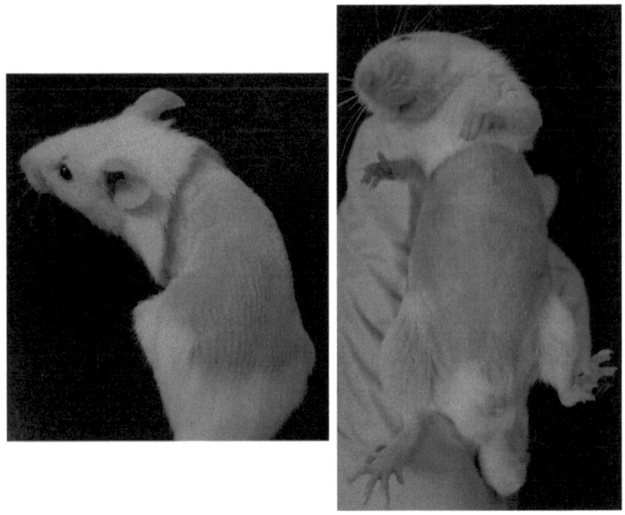

Fig. 1 A mouse shaved whole trunk in preparation for the skin assay. This process should be completed at least the day before carrying out the assay

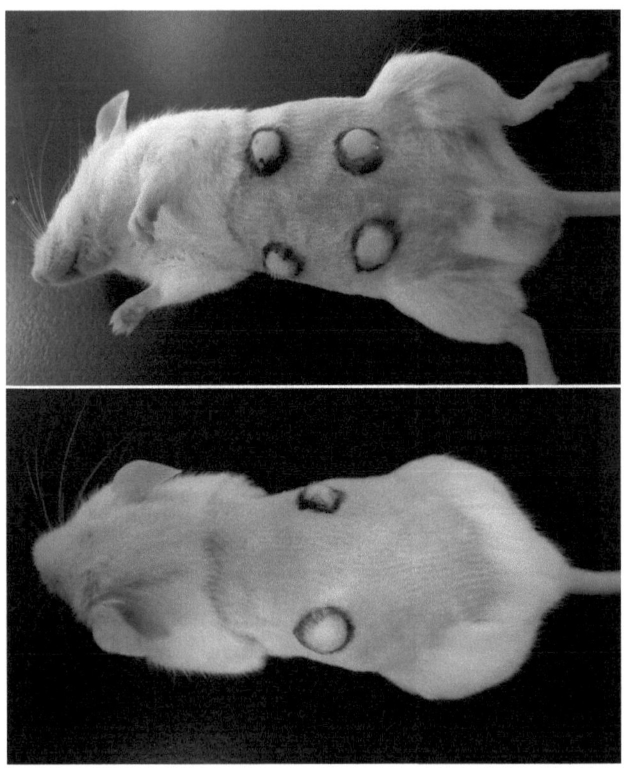

Fig. 2 50 μL of stimulants injected intradermally at six sites. The blebs formed by the injection are marked and serve as a guide for applying the skin punch

3.1.2 Extracting MPO from Skin Samples

Keep samples on ice as much as possible during this procedure.

1. Add 1 mL of MPO buffer to each of the 5 mL tubes containing skin samples.
2. Homogenize tissues using the Polytron tissue homogenizer by placing the tip of the homogenizer at the bottom of the 5 mL tube to tear the skin sample (*see* **Note 9**).
3. Maintain the samples in an ice cold bath and sonicate the tissue, adjusting the power of the sonicator to avoid inducing bubbles (*see* **Notes 10** and **11**).
4. Freeze (−80 °C) and then thaw the samples.
5. Repeat **step 3** and **4**, two more times for each sample.
6. Centrifuge the samples at 1,200 × *g* for 15 min at 4 °C to pellet cell debris.
7. Collect the supernatants (avoiding the whitish material floating on top of the sample) for measurement of MPO activity (*see* **Note 12**).

3.1.3 Measurement of MPO Activity (See Note 13)

1. Add supernatants to a 96-well plate in triplicates (25 μL/well).
2. Add 25 μL of Assay buffer to each well.
3. Start the reaction by adding 200 μL/well of the Development solution.
4. Read the plate at 450 nm every 15 s for 10 min.
5. Collect raw OD at final reading (10 min) for analysis. Omit inconsistent values that can be caused by bubbles.
6. Calculate the delta OD/min using the values of time 0 and 2:20 min.
7. Measure the protein concentration in the sample using a BCA kit according to manufacturer's instructions.
8. Normalize MPO activity to the initial weight of the sample or protein concentration of the extracted sample (*see* **Note 14** and Fig. 3).

3.2 Peritoneal Assay

1. Inject the appropriate amount of irritants (e.g., silica crystals) in 100–1,000 μL PBS into the peritoneum. Inject a control mouse group with the same volume of PBS only (*see* **Note 15**).
2. 4–24 h after injection, euthanize the mice (*see* **Note 16**).
3. Place the mouse on its back and make an incision of the abdominal skin to expose the peritoneal membrane without puncturing it.
4. Inject 6 mL of lavage solution with a 27G needle attached to a syringe (*see* Fig. 4).
5. Pin down the forefeet to the lid of a polystyrene box and shake the distended peritoneal cavity ~100 times.

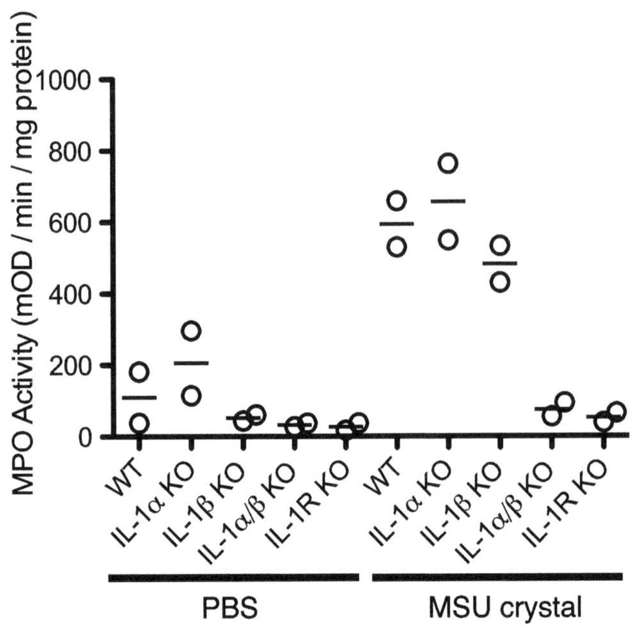

Fig. 3 Skin assay in IL-1 or IL-1 receptor deficient mice. 0.8 mg of monosodium urate crystal (MSU) or vehicle (PBS) alone was injected intradermally in wild type, IL-1α-deficient, IL-1β-deficient, IL-1 α/β-deficient, or IL-1 receptor-deficient mice. 16 h later, MPO activity of the injected areas was measured

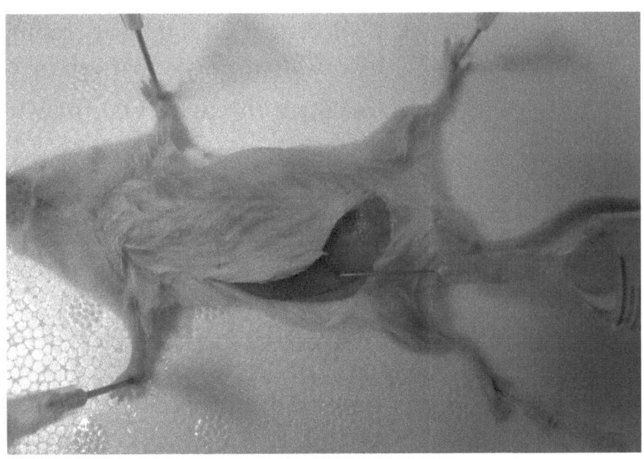

Fig. 4 The tip of a 27G needle connected to a 10 mL syringe containing 6 mL of lavage solution is placed at the exposed peritoneal membrane

6. Collect the lavage with an 18G needle attached to a 10 mL syringe (*see* **Note 17**).
7. Add 500 μL of the lavage fluid into a 1.5 mL tube and centrifuge at 300 × *g* for 5 min at 4 °C (*see* **Note 18**).

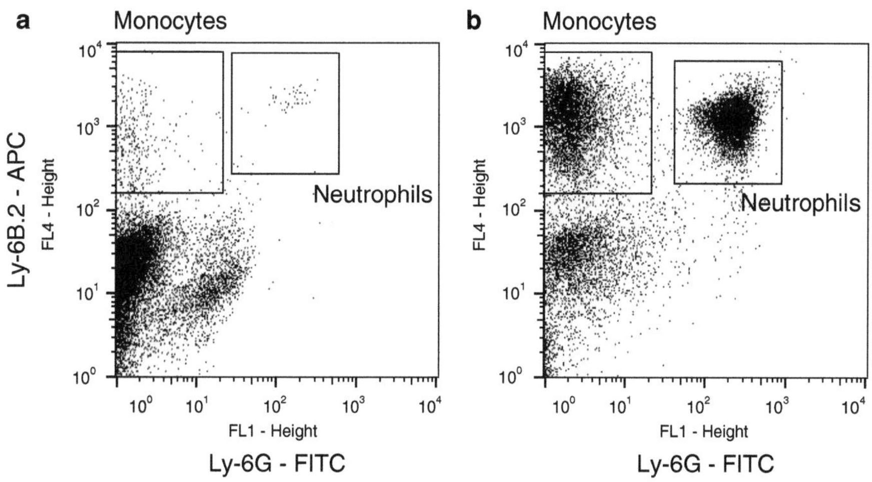

Fig. 5 Representative dot plots of Ly-6G and Ly-6B.2 expression on peritoneal exudate cells in C57BL/6 mice injected 16 h earlier with PBS (**a**) or cholesterol crystals (**b**). The Ly-6G$^+$Ly-6B.2$^+$ gate represents neutrophils and the Ly-6G$^-$ Ly-6B.2$^+$ gate represents monocytes

8. Aspirate the supernatant, resuspend the pellet in 50 μL of staining solution containing 1 μL of Ly-6G-FITC and 1 μL of Ly-6B.2-biotin in the presence of 1 μg of anti FcγRIIB/III receptor antibody, and incubate at 4 °C for 15 min.
9. Add 1 mL of staining solution and centrifuge 300 × *g* for 5 min at 4 °C.
10. Aspirate the supernatant, resuspend the pellet in 50 μL of staining solution containing 1 μL of 7-AAD stock solution (1 mg/mL) and 1 μL of streptavidin–APC, and incubate at 4 °C for 5 min.
11. Add 1 mL of staining solution and centrifuge at 300 × *g* for 5 min at 4 °C.
12. Aspirate the supernatant and resuspend the pellet in 500 μL of staining solution.
13. Determine the total number of neutrophils and monocytes in the peritoneal exudate cells by counting the number of Ly-6G$^+$Ly-6B.2$^+$ (neutrophil population) and Ly-6G$^-$Ly-6B.2$^+$ (monocyte population) cells in 100 μL of stained cells using a flow cytometer with an automatic sampler. We acquire data by CellQuest software and analyze by FlowJo software (*see* Fig. 5).
14. Calculate the total number of the neutrophils or monocytes using the following formula:

Neutrophil or monocyte count in 100 μL of stained cells × [6 mL (volume of lavage fluid injected)/100 μL (cell number count)] (*see* **Note 19** and Fig. 6).

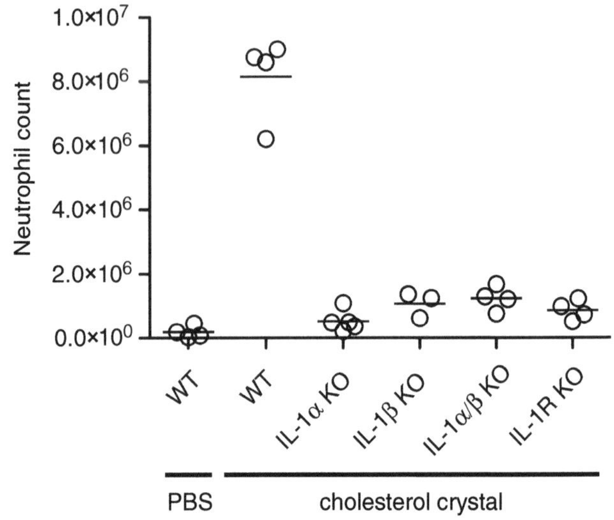

Fig. 6 Peritoneal assay in IL-1 or IL-1 receptor deficient mice. 2 mg of cholesterol crystals or vehicle (PBS) alone was injected intraperitoneally in wild type, IL-1α deficient, IL-1β deficient, IL-1 α/β deficient, or IL-1 receptor deficient mice. 16 h later, the total neutrophil count in the peritoneal cavity was measured

4 Notes

1. Add EDTA from a stock solution (e.g., 1 M) rather than powder, then titrate to pH 6.0; EDTA needs a pH ~8.0 to completely dissolve.
2. This is NOT the *O*-dianisidine (Sigma, D9143), which is not soluble in 50 mM Na_2HPO_4, pH 5.4.
3. Clone 7/4 recognizes Ly-6B.2 alloantigens which are expressed in monocytes and neutrophils [12]. This alloantigen is positive in C57BL/6, DBA, and 129J but negative in Balb/c and CBA mouse strains [13]. In Ly-6B.2 absent strains, Ly-6C can be used as an alternative to Ly-6B.2, which is also expressed on neutrophils and monocytes (mAb for Ly-6C, clone AL-21, #561237, BD Bioscience).
4. The shaving procedure sometimes induces scratching injuries on the skin which can result in a significant increase in MPO activity. Therefore, it is important to shave the skin at least 1 day before the skin assay.
5. We usually put 1–2 pieces of Kim wipes into the 50 mL conical tube and add 2–3 drops of Isoflurane. Place the mouse's nose into the tube to allow breathing of the evaporated Isoflurane. Once the ventilation frequency decreases, take the anesthetized mouse out of the tube and proceed to injections. Continue to monitor the mouse during the procedure to make sure that it remains anesthetized.

6. Try not to inject subcutaneously. It is better to avoid the dorsal side area close to the hind legs, as this area yields significantly lower MPO activity than the other areas. Also do not use the area where a skin wound is found. Shaving the fur may create scratches on the skin, which can dramatically increase the MPO activity of the area.
7. The elevation of MPO activity can be detected at 4 h after the injection of particulate irritants such as silica crystal. The assay duration should be determined according to the design and results of preliminary experiments. We usually pick 4 and 16 h for the evaluation time points, which represent the earliest and the highest MPO activity, respectively.
8. Put the skin punch on the injection bleb guided by the marked outline, and rotate it while gently adding pressure to cut the skin. If this procedure does not cut through the entire dermis, cut the attached skin using scissors according to the marking of the skin punch. When the skin is inflamed in response to the stimulant, the amount of subcutaneous tissue may be much bigger than the non-inflamed control skin sample.
9. Some fibrous tissue can become stuck between the blade and the housing of the homogenizer. Try not to leave any of the fibrous tissue on the homogenizer and clean the homogenizer between samples.
10. The sonication procedure can raise the temperature of the sample. It is better to place the sample in a water bath with plenty of ice and periodically touch the tube to make sure that the sample's temperature remains cold.
11. Sonicating the sample with too much power results in bubble formation in the sample. The sonication power can not be transmitted through the fluid in the presence of bubbles. If bubbles are introduced during the procedure, it is best to wait ~1 min for bubbles to settle, then restart the sonication procedure with lower output power.
12. The whitish material likely represents lipids. It is best to eliminate it as it does not dissolve in the solution and may interfere with the assay, which measures the absorbance of 450 nm.
13. Conjugate and develop one plate at a time.
14. In Fig. 3, the final MPO activity was compensated with the protein concentration of the assayed solution of the extracted skin. For a point of MSU stimulated WT mouse, the raw MPO activity was measured as 40.4 mOD/min. The protein concentration was 2.45 mg/mL. The correction was calculated as follows: The correlated MPO activity of the sample, 659 (mOD/min/mg) = 40.4/(0.025 (mL) × 2.45 (mg/mL)). Note that we used 25 μL of the skin extract for the assay.

15. It is better to check the dose responses and determine the amount of the stimulants aiming at the middle of the dose response reactions. In Fig. 3, the amount of the MSU crystal used for the assay was 0.8 mg in 50 μL PBS. We found this amount to be appropriate as 1.6 mg of MSU crystals per site yielded 50 % higher MPO activity.
16. The assay duration should be determined according to the design and results of preliminary experiments. A substantial number of neutrophil recruitment can be seen at 4 h after the injection of particulate irritants such as silica crystal. We usually pick 4 and 16–20 h for the evaluation time points which represent the earliest and the highest neutrophilic infiltration, respectively.
17. We usually draw back 4–5 mL of the lavage fluid although only 0.5 mL of lavage is required for the following procedures. When one uses older mice, the peritoneal fat tissue may block the tip of the needle which may interfere with the aspiration procedure. To avoid this from happening, insert the needle from the lower part of the peritoneal cavity. This allows easy access to the peritoneal cavity. Place the tip of the needle between the peritoneal wall and the lateral side of the liver, as this prevents fat tissue blocking the needle during retrieval of the peritoneal fluid. It is also important to take care not to puncture the intestine or liver.
18. The rest of the lavage fluid can be centrifuged and stored for further analyses, i.e., cytokine measurements.
19. Cell count may be calculated by determining the ratio of fluorescent count beads (e.g., Flow Cytometry Absolute Count Standard™, Bangs Laboratories, Inc., Fishers, IN) to the cells. Another way to quantify the number of cells is as follows: (1) Count the number of cells in the peritoneal exudate using a cell counting chamber (cells/mL). (2) The total number of neutrophils recruited to the peritoneal cavity can be calculated by multiplying the number of cells from the chamber slide (cells/mL) by 6 mL (volume of lavage injection) and the percentage of neutrophils or monocytes shown as the result of flow cytometry (*see* Fig. 5).

Acknowledgment

We thank Natalie Olive for critical reading of the manuscript.

References

1. Dinarello CA (2009) Immunological and inflammatory functions of the interleukin-1 family. Annu Rev Immunol 27:519–550. doi:10.1146/annurev.immunol.021908.132612
2. Hoffman HM, Rosengren S, Boyle DL, Cho JY, Nayar J, Mueller JL, Anderson JP, Wanderer AA, Firestein GS (2004) Prevention of cold-associated acute inflammation in familial cold autoinflammatory syndrome by interleukin-1 receptor antagonist. Lancet 364(9447):1779–1785. doi:S0140673604174011 [pii] 10.1016/S0140-6736(04)17401-1
3. Hawkins PN, Lachmann HJ, Aganna E, McDermott MF (2004) Spectrum of clinical features in Muckle-Wells syndrome and response to anakinra. Arthritis Rheum 50(2):607–612. doi:10.1002/art.20033
4. Hoffman HM, Mueller JL, Broide DH, Wanderer AA, Kolodner RD (2001) Mutation of a new gene encoding a putative pyrin-like protein causes familial cold autoinflammatory syndrome and Muckle-Wells syndrome. Nat Genet 29(3):301–305. doi:10.1038/ng756 ng756 [pii]
5. Kanazawa N, Furukawa F (2007) Autoinflammatory syndromes with a dermatological perspective. J Dermatol 34(9):601–618. doi:JDE342 [pii] 10.1111/j.1346-8138.2007.00342.x
6. Kono H, Rock KL (2008) How dying cells alert the immune system to danger. Nat Rev Immunol 8(4):279–289. doi:nri2215 [pii] 10.1038/nri2215
7. Rock KL, Latz E, Ontiveros F, Kono H (2010) The sterile inflammatory response. Annu Rev Immunol 28:321–342. doi:10.1146/annurev-immunol-030409-101311
8. McCarty DJ, Hollander JL (1961) Identification of urate crystals in gouty synovial fluid. Ann Intern Med 54:452
9. So A, De Smedt T, Revaz S, Tschopp J (2007) A pilot study of IL-1 inhibition by anakinra in acute gout. Arthritis Res Ther 9(2):R28. doi:ar2143 [pii] 10.1186/ar2143
10. Mullane KM, Kraemer R, Smith B (1985) Myeloperoxidase activity as a quantitative assessment of neutrophil infiltration into ischemic myocardium. J Pharmacol Methods 14(3):157–167
11. Kono H, Karmarkar D, Iwakura Y, Rock KL (2010) Identification of the cellular sensor that stimulates the inflammatory response to sterile cell death. J Immunol 184(8):4470–4478. doi:jimmunol.0902485 [pii] 10.4049/jimmunol.0902485
12. Rosas M, Thomas B, Stacey M, Gordon S, Taylor PR (2010) The myeloid 7/4-antigen defines recently generated inflammatory macrophages and is synonymous with Ly-6B. J Leukoc Biol 88(1):169–180. doi:jlb.0809548 [pii] 10.1189/jlb.0809548
13. Hirsch S, Gordon S (1983) Polymorphic expression of a neutrophil differentiation antigen revealed by monoclonal antibody 7/4. Immunogenetics 18(3):229–239

INDEX

Christine M. De Nardo and Eicke Latz (eds.), *The Inflammasome: Methods and Protocols*, Methods in Molecular Biology, vol. 1040, DOI 10.1007/978-1-62703-523-1, © Springer Science+Business Media New York 2013

R

S

T

V

W

MIX
Papier aus verantwortungsvollen Quellen
Paper from responsible sources
FSC® C105338

If you have any concerns about our products,
you can contact us on
ProductSafety@springernature.com

In case Publisher is established outside the EU,
the EU authorized representative is:
Springer Nature Customer Service Center GmbH
Europaplatz 3, 69115 Heidelberg, Germany

Printed by Libri Plureos GmbH
in Hamburg, Germany